THE SPIRIT OF THE RAINFOREST

DR ROSA VÁSQUEZ ESPINOZA

THE
SPIRIT
OF THE
RAINFOREST

How Indigenous Wisdom and
Scientific Curiosity Reconnects
Us to the Natural World

First published in Great Britain in 2025 by Gaia, an imprint of
Octopus Publishing Group Ltd
Carmelite House
50 Victoria Embankment
London EC4Y 0DZ
www.octopusbooks.co.uk

An Hachette UK Company
www.hachette.co.uk

The authorized representative in the EEA is Hachette Ireland, 8 Castlecourt Centre, Dublin 15, D15 XTP3, Ireland (email: info@hbgi.ie)

Copyright © Dr Rosa Vásquez Espinoza 2025

Distributed in the US by Hachette Book Group
1290 Avenue of the Americas, 4th and 5th Floors
New York, NY 10104

Distributed in Canada by Canadian Manda Group
664 Annette St., Toronto, Ontario, Canada M6S 2C8

All rights reserved. No part of this work may be reproduced or utilized in any form or by any means, electronic or mechanical, including photocopying, recording or by any information storage and retrieval system, without the prior written permission of the publisher.

Hardback ISBN: 978-1-85675-556-6
Trade Paperback ISBN: 978-1-85675-557-3
eISBN: 978-1-85675-559-7

A CIP catalogue record for this book is available from the British Library.

Typeset in 11.75/18.5pt Heldane Text by Six Red Marbles UK, Thetford, Norfolk

Printed and bound in Great Britain.

13 5 7 9 10 8 6 4 2

This FSC® label means that materials used for the product have been responsibly sourced.

I dedicate this book to my grandmother, who dreamed of becoming a doctor. She was the first healer I met.

CONTENTS

Introduction	1
Chapter 1: Pachamama	9
Chapter 2: A River That Boils	29
Chapter 3: Life in the Water	49
Chapter 4: Evil Spirits	73
Chapter 5: Stingless Bees	97
Chapter 6: Ayahuasca	119
Chapter 7: What Lurks Beneath the Murky Waters	139
Chapter 8: Living Fossils	163
Chapter 9: Living in the Clouds	183
Chapter 10: Ancestral Civilizations	205
Chapter 11: In the Eyes of the Jaguar	231
Chapter 12: Life that Glows in the Dark	251
Epilogue	275
Acknowledgments	279
Picture Credits	283
Endnotes	285

Before you step into the jungle, there are a few things you need to know.

INTRODUCTION

– What are you most afraid of in the jungle?

We were coming to the end of the interview and it was time for the rapid-fire questions: quick question, quick answer, no follow-up, no background. It was one of the most insightful conversations I've had with a journalist.

As a scientist and National Geographic Explorer, I often get asked about the Amazon Rainforest. Conversations naturally fall into the topics of biodiversity, climate change, natural medicines, and travel— but it is when people start to ask the deeper questions that things get more interesting.

The Amazon is the most biodiverse region on our planet: a living rainforest with thousands of species stretching as far as the eye can see. It is filled with scents, sounds, textures, hues, and life forms that embody the purest manifestation of nature. The raw and enchanting beauty of the rainforest contrasts with its extremely demanding conditions, testing the resilience of anyone who ventures into it with its oppressive humidity, scorching temperatures, impenetrable canopies, and deadly creatures that lurk in its shadows.

Yet, the deeper I go into the Amazon jungle, the more I dive into exploring its hidden worlds: the age-old stories passed down through generations; the legends that characterize the tribal areas, and the wisdom of its indigenous people.

It all started with my grandmother, a traditional healer from Perú.

When I was five, I remember strolling through the luscious rainforest in my favorite pink dress, proudly holding my grandmother's hand. The air smelled of citrus and sweet fruits, the expansive rivers intertwined with endless greens, and the stiffening humidity made my hair curl. Summertime had arrived, and I wore a big smile on my face as I followed my grandmother around the rainforest, fascinated by all I saw. It's one of the fondest memories I have from my childhood; we were joint explorers, uncovering forgotten worlds and uncharted territories together. She was also my teacher. While she couldn't name any scientific classifications of plants, nor explain their biological processes, she did know how to combine a flower's roots with a tree's bark to stop a chest infection, or how to tap in to the ancestral wisdom of birds. Ultimately, she taught me how to connect with the deepest lessons from the Pachamama[1] (Mother Earth). Through respect and knowledge of the natural resources available to us, she shaped my vision of the rainforest. Plants and shrubs had the potential to provide a cure or a poison, and it was through the traditional wisdom of this place and its people that my grandma taught me the difference.

– *Todo está vivo* – she'd say. Everything is alive. Including what conventional biology says is not. Her teachings irrevocably marked my indigenous upbringing and also my subsequent journey as a scientist.

A subtle cough drew me back to the video call as my mind refocused on the question.

– *What am I most afraid of? Well, the evil spirits of the jungle* – I replied decisively, ready for the next question. I had my sleeves rolled up, and I leaned slightly toward the screen. I was ready.

I realized I had startled the journalist with my response, as his pupils dilated and his eyebrows moved inward. He was trying to make sense of what I had said.

– *OK, I must step away from protocol here . . . what the hell do you mean?*

– La curiosidad mató al gato* – I thought to myself. I saw the clock and knew I would end up running late, but I couldn't resist his curiosity. I took the bait.

– *The* chullachaki,† *of course* – I said with confidence, almost assuming everyone would understand.

His eyebrows reminded me that most of the world didn't.

– *There are spirits in the rainforest, and many of them are good, but the* chullachaki *is one to stay away from. People use Amazonian tobacco, garlic, and salt to keep it away. Legend says that if your tobacco is consumed by the end of the night, and you didn't touch it once, then the*

---

* *La curiosidad mató al gato* means "curiosity killed the cat"—a proverb that implies that pursuing certain curiosities can sometimes lead to danger or misfortune.

† A mythical creature that is often described as a small, ugly spirit with one leg shorter than the other, and one foot larger than the other. It is regarded as a guardian of the land that punishes those who act unwisely or break a taboo.

chullachaki *had been close but accepted the* ofrenda* *of the tobacco and left you unscathed.*

He took a minute to process this, and part of me was expecting mockery.

I used to care if people questioned or laughed at my view of the world. I used to long for people to see me as a 'serious scientist'—to believe that my background needed to be hidden. My indigenous upbringing in the Amazon and Andes of Perú, my knowledge of traditional medicine, and my years as a professional dancer, meant I needed to act seriously as a Ph.D. scientist in the US in order to be taken seriously. I'm not sure when that changed. But embracing the idea that we do not owe people explanations now fills me with courage. So, instead of getting an unsettled feeling in my stomach in anticipation of his response, I simply smiled and waited.

– *Have you encountered it?* – he asked in a very serious tone, his eyes fixed in fascination.

– *No . . . not yet*—knock on wood—*but I have heard first-hand experiences from many of my field collaborators, and they are chilling.*

The collaborators on my expeditions are strong Amazonian *Apus*† (masters) who have faced some of the most ferocious predators of the rainforest. And yet, if you ask them what they are most afraid of, most will undoubtedly say with a hint of fear in their voice that it's the evil

---

* *Ofrenda* means offering. In Peruvian indigenous cultures across the Amazon and the Andes, it is very common to give offerings to the land to honor and show respect to the spirits that reside in the natural world.
† *Apu* is a Quechua word to describe a natural spirit, a sacred mountain, or, more colloquially, an important and respected leader.

spirits of the jungle. They do not mess around, so the *Apus* always carry protection against these creatures, just in case. Garlic, salt, and Amazonian tobacco.

Upon leaving Perú to pursue my studies and traveling to new continents, I soon learned that the stories and ancient wisdom I had grown up with were not common to everyone. In fact, they were far from ordinary. This book is about the stories I've encountered in the Amazon Rainforest and the fascinating discoveries that challenge science as we know it. Throughout the next 12 chapters, I'll take you on a journey through my adventures—with insights from the Andes, discoveries made in the field, and stories recounted from *Apus*, including my grandmother—all in my quest to reconnect with the natural world. These are experiences, anecdotes, and lessons that helped me attune to the spirit and beauty of nature—something we can all learn to do if we are willing to observe and listen to the forest.

I write from a personal standpoint about my Amazon experiences, and I will not reveal the precise locations of our expeditions to protect these ecosystems and communities from an influx of visitors. But, before we begin, I want to express my deep gratitude to all my *Apus* and mentors, including my grandmother, who have shown up in my journey in so many unexpected ways. I thank them for allowing me to enter their sacred spaces and expansive worlds, helping me to discover new nuances of life and to dive deeply into the indigenous knowledge of my ancestors.

– *Do you believe in spiritualism, then?* the journalist continued in this now-extended interview, as my eyes flicked to the clock for a second time.

A deep question with a simple answer.

I find it impossible to separate my scientific explorations from inherent spirituality—whether I am trekking in the heart of the Amazon Rainforest or climbing the high Andes of Perú. As scientists, we are trained to look at data, hard and irrefutable evidence, yet we are also encouraged to be creative—to think of new theories, new possibilities, new experimental designs that prove or disprove ideas. And in this way, we must be attuned to the beliefs, practices, and taboos—the culture—at the heart of our scientific endeavors. Spiritualism and science can work together, and this approach marks my path and worldview. It has also inspired the 'pharmacopeia' note at the beginning of each chapter, in which I highlight a unique Amazonian organism from both traditional and scientific perspectives, in a way that mirrors how I explore the natural world.

For as long as the Amazon and the Andes have existed, nature and culture have prevailed in a complex symbiotic dance. We may not have Inca or Amazonian leaders reigning our lands or waters anymore, but their beliefs live on today. From their devotions to the sun and their offerings to the mountains, to their knowledge of medicinal plants and their mastery of balance within the spirit, the indigenous voices of the Amazon understand our deep interconnectedness with nature in ways that we do not. But I'm on a mission to find out. Reconnecting with the natural world is not only a journey into our own humanity; it is key to sustaining the biodiversity and cultures of our planet.

Growing up in Perú, I remember how my uncle in Cuzco always used to pour a bit of his drink of pisco[2] onto the soil below our feet before drinking any himself.

– It's for the Pachamama – he used to say. *She drinks first.*

INTRODUCTION

In the ancestral territories of Perú, nature is seen not just as a backdrop but as a living presence, revered and always there. This is a worldview that both respects and considers Mother Earth before anything else, and it's one that we would be wise to learn from.

So hear, hear, for our Pachamama, and let's begin our journey . . .

1

PACHAMAMA

Palo Santo

SCIENTIFIC NAME: *Bursera graveolens*

TRADITIONAL NAME: Palo Santo

ORIGIN: Native to the tropical forests of South America, including Perú

TRADITIONAL USES: Palo Santo or "Holy Wood" is a powerful medicinal tree revered for its aromatic wood and ancestral uses by communities in the Amazon Rainforest and Andean mountains. Traditionally, local communities harvest fallen branches or trees that have died naturally and have been resting for four to eight years. This waiting period allows the aromatic properties to fully develop. In sacred rituals and ceremonies known as *sahumerios*, the wood is burned, releasing a unique aroma resembling a mix of pine, citrus, and mint, with a sweet and woody fragrance. These *sahumerios* are used to cleanse energies, bring about spiritual purification, and ward off evil spirits. Many burn Palo Santo sticks as incense nowadays. Additionally, Palo Santo essential oil can be harvested and used topically, as it has calming and anti-inflammatory properties.

SCIENTIFIC INFORMATION: Palo Santo is rich in aromatic compounds such as limonene and α-terpineol, among other terpenes and essential oils, which contribute to its uses in traditional medicine. The tree typically grows up to 4–10 meters (13–33 feet), although it has been recorded that it can reach up to 20 meters (65 feet). Sustainable harvesting is essential to protect this valuable natural resource for future generations.

It was 4:45 in the morning, and I couldn't force myself back to sleep. The electronic alarm hadn't gone off yet, but I could hear the faint and distant crow of a rooster. My favorite natural alarm. I smiled as I turned my body around, slowly rubbing my eyes as they opened to greet the dawn of the Amazon Rainforest.

I could hear tiny frogs jumping outside, gently slapping their feet against our wooden hut; their cheeky movements were subtle, but the sound echoed through the wood, as though thick drops of rain bounced playfully against our walls. A family of insects flew through the air, whirring loudly, pulling my attention away from the frogs. Everyone in the hut was fast asleep. Our day wouldn't start for at least two hours, but I was already impatient to see the morning haze rise above the mountains that surrounded us. Tall and imposing against a half-moon, the mountains were watching us arrive, curious about our intentions. The territory in the Southern Amazon was new to me, and I was new to it, and so it felt like a morning introduction was due.

I squinted, scanning the floor below me for any unexpected visitors. In the rainforest, I've encountered all sorts of creatures in our huts: lizards hiding in my shoes, frogs hanging on walls, snakes entwined in clothes. There was even a time when a tapir—one of the largest mammals in the Amazon, also known as a mountain cow—licked my friend's face, like a dog waking up its owner. So yes, it pays to take a careful look.

No unusual guests were wandering about this morning, and I sensed my secret disappointment. But as I began to dress, I noticed a trail of ants creeping through my backpack, heading toward my socks and the legs of my bed. I quickly inspected my cargo pants—all clear. My shirt? Clear too. Checking my long black hair, I shook it vigorously, otherwise unable to identify any black ants against my pitch-black mane. Thankfully, all clear. I let out a sigh of relief—though the paranoia left me feeling itchy.

Resisting the temptation to unnecessarily scratch, I grabbed my headlamp, threw on a sweater over my shirt, and kept my hair down for extra warmth. I could hear the faint whistle of wind outside, and I knew early mornings in the Amazon were often breezy. I could almost hear my grandmother telling me *que no te de viento* (don't let the wind get you), as the ancestral belief in South America reminds us that a gust of wind can cause colds or muscle aches.

I made my way out of the hut and entered the crisp but typically humid morning of the Amazon Rainforest. Chirping, squeaking, clicking: the sounds had intensified. The jungle was as alive as ever, welcoming me home. It was the third time I had entered the Amazon Rainforest that year, yet every time felt like the first. I walked a few steps around our hut, counting every scent I could detect as I closed my eyes and breathed in the rainforest: cinnamon, *maracuya* (passion fruit), cedar, and a slight hint of eucalyptus. I could even smell the earthy, rich and moist aroma of wet soil known as "petrichor." This complex mixture of scents is primarily composed of plant oils and *geosmin*—a natural molecule produced by a soil microbe called *Streptomyces*. This microbe plays an essential role in decomposing organic material to maintain the health of our soils, and it also has the extraordinary ability to produce lifesaving antibiotics like tetracycline and vancomycin. The word "petrichor" is derived from ancient Greek, and means "the ethereal fluid that is the blood of the gods." Musky, fresh, slightly sweet, and personally addictive. The intense rain that had showered us the night before was suddenly worth it. No combination of synthetic perfumes could ever surpass this natural morning fragrance.

I descended the hill, moving away from the towering dark trees and the dense flora behind our cabin. As I made my way toward the community's center, the twigs crackled below my feet and a sliver of sunlight broke through the clouds, reflecting against the forest floor. I walked past numerous huts, some as quiet as my own, others echoing with the clattering and clanging of pots inside.

A couple of children played nearby, kicking around a partially inflated ball; their laughter filled the air as playful small dogs scurried behind them, chasing the ball. Several people were sitting outside their homes: some crossing legs next to small fires, others chatting softly while preparing leaves and peeling yucca. It was the harmonious and hypnotizing morning routine of the Amazon.

I was staying with the Ashaninka community, one of the largest indigenous groups in the Peruvian Amazon. Known for their rich history of ancestral traditions, they possess a profound ecological knowledge of the rainforest I had come to explore. We were conducting a reconnaissance of the area, which has an extreme mix of reduced oxygen, lowered temperatures and high humidity, and unique native wildlife; these life forms have adapted their physiology, chemistry, and DNA to live 1,500 meters (5,000 feet) above sea level in one of the last ecosystems of high-altitude rainforest left on our planet. I also wanted to learn more about the Ashaninka people's ancestral plant- and animal-based medicines, which date back centuries. Most Ashaninka knowledge is preserved orally but, with the increasing threats that the Amazon is facing, their wisdom runs the risk of disappearing, along with the rich biodiversity and palpable culture.

– Kitáíteri! *(Good morning)* – greeted me on my arrival. Micaela, one of the kindest women I have ever met, with cooking skills like a goddess, was skillfully preparing breakfast for her family while also scattering seeds and leftover fruit for her chickens, her long black hair tied in a bun. She was a prominent female leader in the community. Her husband, Pascual, the *Apu* (leader) of the community, was nearby, washing his face in a bucket with fresh water drawn from the river. He was a man of medium stature, with deep, soul-piercing dark eyes, and a kind, reassuring expression. Barefoot, he wore his *cushma*—a striking, sleeveless, bright orange tunic adorned with geometric patterns that reflected his cultural inheritance and leadership role. These patterns, created using natural dyes, demonstrate the stories and spiritual beliefs of the Ashaninka people. Some *cushmas* may display undulating waves that represent water, while others may showcase triangular peaks that reflect the spirit of the mountains, or circles that emblemize the sun. *Cushmas* are simple yet profoundly symbolic; they embody the Ashaninka's self-sufficiency, their deep connection to the land, and their commitment to traditional practices and ecological knowledge.

– Kitáíteri doctora! – Pascual approached eagerly, welcoming me with a warm hug, asking if our team had managed to sleep through the heavy rain of the past night. The Ashaninka are renowned for their kindness and hospitality toward guests, with a strong emphasis on collective well-being. Pascual and the community were excited to host us—a team of scientists and storytellers—as we came to document the extraordinary biodiversity of their home, and discover the profound cultural traditions that seek to preserve the rainforest. In our short time here, they already felt like family.

The Ashaninka people pursue a philosophy of "living beautifully" or "good living," which is known in various native languages, including *kametsa asaiki* (in Ashaninka), *jakon jaki* (in Shipibo-Conibo), and *sumak kawsay* (in Quechua). This philosophy is prevalent across Amazonian and Andean cultures. It is fundamental to the Amazonian cosmovision or worldview—an indigenous way of seeing the world, a paradigm that embodies living harmoniously with nature. These communities see nature as an integral part of human coexistence, not as a separate entity, and their worldview involves living interconnectedly with the rivers, lands, forests, animals, and plants, taking only what is needed—just like my grandmother taught me. It is a way of living that means reciprocating with the Earth, constantly maintaining balance throughout time. It also means honoring the invisible spiritual bond we have with the natural world. Beautiful living, *living beautifully*, is not an individual pursuit, but rather a collective endeavor, embodying the essence of the Pachamama (see page 285).

Pascual and I made our way down to the lower part of the hill, heading toward a large field at the community's heart. The space was used for reunions, rituals, dancing, and simply admiring the environment they call home. For a few minutes, we shared the beauty of the rainforest in silence. Then Pascual noticed that my eyes were fixed on a tall mountain in the distance, and he began to recount one of the most captivating stories of creation I had ever heard.

– *That is the* Avireri, doctora – the *Apu* said, his voice deep with wisdom.

I stood silent, admiring the infinite shades of green in the mountain's canopy that were sporadically reflecting the sparkling glares of the sun.

– *In the beginning of time, everything was dark. Like when there is no moon, you know?*

I nodded slowly, and he continued.

— *Everything was cold and frightening. Everything was chaos. Everything was human. All we see around us, including birds, fish, monkeys, trees, flowers, streams, and hills, they were all human before taking the form that they have today; whether that be animal, floral or landscape.*

— *The early humans lived in perpetual darkness and disorder, in a shapeless world. Like if mountains and rivers were to disappear, you know? But humans were not completely alone: invisible beings dwelled among them.*

Pascual paused, then continued. *The spirits taught humans basic survival skills, imparted cultural practices, and molded human identity and character. Benevolent spirits, known as* amatsénka, *imparted good values like compassion, generosity, and kindness, inspiring humans to be good people. However, equally present were bad, malevolent spirits, called* kamári, *that rejoiced in causing harm, encouraging jealousy, aggression and selfishness, and bringing about fear and chaos in the early world.*

— *But then the* Avireri *emerged to separate day from night.* Kashirí, *the moon, and* Pavá, *the sun, male and female, ascended to the sky, giving birth to the dry and rainy seasons, while creating beautiful music for one another.*

Pascual's *cushma* gently billowed in the subtle breeze, revealing the bird drawings in dark ink at its hem.

He went on – *Then lands and oceans were created, and some humans were transformed into animals and plants that turned our world into a beautiful one. But the spirits that inhabited the world at the beginning of time are still around today. The good spirits,* amatsénka, *sometimes camouflage themselves as magnificent animals with extraordinary powers that we humans can only dream of, like the hummingbird that soars high and far. They materialize into animals to experience the beauty that we still have left in the world, and they serve as symbols of good omens and protection.*

– We Ashaninkas also believe that good spirits transmit us important medicinal knowledge. They talk to us through dreams to share what plants or leaves we must use to combat illnesses or ward off the kamári, *evil spirits, that are still around us, causing disease, unfairness, and sadness today. It is our Ashaninka belief that nothing is ever truly destroyed; everything is transformed. And our great hero of transformation, the* Avireri – he said, proudly gesturing toward the mountain in front of us – *is who transformed it all.*

I took a deep breath as I absorbed his words. The cultural significance of these lands felt immense; as the story unfolded, the mountains seemed to grow more majestic, and my commitment to respectful exploration deepened.

– My grandfather, a sheripiári *(shaman), used to tell us that the* Avireri *chose these new forms based on the behavior of the early humans* – Pascual continued. *This means that the human essence persisted in the new being after transformation. If the human was mischievous, the transformed being was mischievous. If the human was generous and giving, so it was in its next form of life.*

Pascual explained that if a human was overly violent, the *Avireri* transformed them into wasps. Those caught stealing became *monos choros* (stealing monkeys).[1] Those who prepared the best drinks became native bees that made delicious honey. Tree-cutters were turned into woodpeckers, those who drank excessively were changed into flies who were attracted to fermented beverages, and enemies were morphed into rocks. Legend even has it that an invading ship was transformed into the cliff we see in the Tambo river, and its sailors were perpetually transformed into dozens of ants.

– These transformations mean that when we die, it is only because our life has been taken by the natural world; our spirits continue to live in a river, an animal, a flower, a mountain, a tree, forever interconnected – he said, interlocking his fingers in a symbolic gesture.

Pascual added that his grandfather was now a powerful medicinal tree found deep within the forest. To this day, he provides them with curative leaves and roots that are essential ingredients in their traditional medicines, including some that treat against COVID-19. I couldn't help but wonder what my great-grandmother, my sweet *leona* (lioness), might have become in the Ashaninka realm. Perhaps her nickname became her reality.

– Everything in our world has a spirit and is alive.

Pascual's voice brought me back to the present. He explained that their community seeks the *Avireri*'s permission and blessings before entering the mountains to hunt or forage. They do so through quiet prayer rather than through offerings or rituals. They ask the *Avireri* to protect them as they navigate the forest, feeling their bodies fill with positive energy. Each prayer is individual to the Ashaninka. They are encouraged to speak to the *Avireri* in their own words, expressing their unique emotions, thoughts, and hopes for the journey. This respect for the great transformer is a commitment to reciprocate what they take, and to seek protection when they wander deep into the heart of the Amazon. Sometimes trekking too far into the rainforest leads to encounters with hungry, ferocious wildlife or deadly poisonous species, but if the *Avireri* is sought, these will be peaceful interactions. However, those who disregard the respect for the *Avireri* always face the same fate: none ever returns.

— *So everything on our planet possesses a human-like spirit* — I repeated, as my mind sought to understand such ancient wisdom.

It became clear to me that there was little room for the Pachamama or stories of the *Avireri* in my formal scientific education. In conventional science, mountains do not talk, rivers do not embrace powerful spirits, plants do not share their curative powers through dreams, and animals do not guide you in times of need. There is no way of quantifying or measuring these premises. As such, many would discount these stories as illogical beliefs. They would set them aside, leaving them for indigenous people and self-made gurus who claim to have the solution for each and every problem. This realization led me to understand that the Pachamama transcends scientific inquiry. And yet, the disconnect between ancient wisdom and modern science is accelerating the loss of all that is beautiful in our world.

— *Our entire Pachamama is alive*, doctora — Pascual wisely added.

Somewhere within, my ancestral roots resonated with this truth.

I've always straddled the worlds of spiritualism and science. Growing up in Perú, my grandmother taught me how to cultivate, prepare, and grow remedies right in our backyard. She had learned these techniques from the elders in her community many decades before. She taught me how to talk to and tap in to the wisdom of the soil, the flowers, the water, the animals, and the sun. *They can all hear you* — she always said.

Meanwhile, during my years pursuing a scientific career in the Western world, my mentors taught me how to question every piece of evidence and collect tangible data—only relying on what I could irrefutably see and count. They taught me to stick to the parameters of scientific examination, as science, they said, is the highest expression of human intelligence; it is the only explanation of the natural world that we can trust.

In biological terms, rocks, mountains, and rivers aren't alive. To be "alive" biologically speaking requires the ability to grow, reproduce, and respond to stimuli in a measurable manner. Since the birth of modern science, a rigorous methodology based on logical analysis and evidence has dominated our quest to understand the natural world. Integrating both experimental and mathematical techniques, skepticism and observation was central to the scientific method of discovery throughout history.

And yet, I see great strength in bringing together my scientific training with my indigenous soul to offer new ways of seeing. Even within scientific disciplines there are different ways to define life. Consider physics, for example; physics teaches us that everything in the universe is composed of particles and fields, and the behavior of these are dictated by the laws of physics, which are based on ever-evolving mathematical models used to approximate the laws of nature. These laws are applied uniformly to all particles, whether considered living or non-living entities. Intrinsically, at the atomic and subatomic levels, there are no differences between the particles that make up our human bodies and those that make up a bed of rocks or water flowing down a river. So, to a biologist, perhaps the lens of the *Avireri* has as much to teach us as the perspective of physics.

Whether viewed through the *Avireri* or the principles of physics, the question arises: can we find a reality where we more fully embrace the ancestral wisdom of our lands? Is there a way to reconnect with our Pachamama, our natural world, by exploring alternative and creative methods of understanding life? Throughout my scientific journey, it is the fundamentals of dance improvisation that have led to my most significant findings and progress. It was through dance that I was inspired to explore the fluidity of molecular movement that led to novel antibiotics; and through the arts I discovered the importance of socio-emotional communication when it comes to influencing policy. It is the questions we pose and the stories we tell that define how we see the world; and it is these ways of seeing that become the catalyst for who we are and how we act. So how do we listen to those who see differently?

It is this very question that keeps me returning to the rainforest. As a scientist, I feel the urgency to document life forms that have historically gone ignored, and yet play a key role in maintaining local ecosystems, even holding the key to discoveries that benefit the rainforest and humankind. Yet, as an indigenous Amazonian-Andean descendant, it is my right and responsibility to show that nature doesn't occur independently of culture and people, and thus, science shouldn't either. It is my upbringing, identity, and closeness to the rainforest that encourage me to integrate modern science with indigenous wisdom, elevating academic and ancestral knowledge. What can we learn from the rainforest to help us find new ways through which we can see nature, inspiring empathy and fueling curiosity? And how can this help us to protect our planet? I am on a mission to find out.

Higher on the horizon, the sun's golden tendrils pierced through the thick canopy. The haze was now clearing, indicating it was time to return to the hut and join my team for a hearty breakfast. I picked up my sandals from the ground, and began my journey back; the crisp sensation of the grass below my feet was refreshing and calming.

As I ascended the hill, my heartbeat grew stronger. I was thinking of the large, warm loaf of bread I planned on dipping in hot coffee and devouring. But, as I passed the huts, I heard the faint sound of someone singing softly in the Ashaninka language and my curiosity got the better of me. As I followed the sound, I peered into a hut and saw Marina, the *Apu*'s mother.

Marina was a beautiful elder Ashaninka woman with mid-length hair as black as night, and tanned skin adorned with joyful wrinkles from a lifetime in the jungle. She wore long heavy necklaces strung with beads of different sizes. The beads accompanied her singing, rattling as she moved her body. Her face was decorated with red paint in geometric shapes, both to shield her from the approaching sun and to express her intentions for the morning. She was wearing a *cushma* of a dark orange hue, with long black curvy lines running along the bottom, reminding me of the underground network of plant roots.

As soon as Marina saw me, her eyes widened and a broad smile lit up her face, calling me to approach. Marina didn't speak Spanish, and she knew my Ashaninka was currently limited to *kitáiteri* (good morning) and *pasonki* (thank you), yet it seemed like she wanted to share her extensive collection of natural remedies with me. She was in the midst of preparing a traditional medicinal ritual that can only take place early in the morning, before the sun has reached its zenith in the sky.

We stood beside her home, sheltered under a beautifully constructed tent made from sturdy tree trunks and lush leaves that provided shade.

Right in the middle, Marina had started a fire that simmered with white embers. An old and blackened cooking pot sat on top of the fire, filled with water that was approaching boiling point. Next to Marina lay a fascinating assortment of rocks, leaves, twigs, and roots, spread out on the bare ground.

Marina hugged me, then guided me around while speaking words in Ashaninka that I couldn't grasp, but which somehow made sense. It seemed like the rocks originated from a special location toward the right side of the mountains. The leaves and roots varied in sizes and color, and while I recognized some, others were entirely new to me.

As I examined her collection more closely, there was one fragrance I instinctively recognized, causing my eyes to light up. Marina sensed I recognized the Palo Santo, and she grabbed more of the aromatic wood from a hidden corner, bringing it closer to me so that I could fully appreciate its rich scent. A woody, resinous tree bark with hints of warm spice and the slight freshness of citrus. The sweet, soft smell of growing up in an Andean-Amazonian rooted household.

Palo Santo is an essential element in various spiritual and energetic ancestral rituals, including the *sahumerio*—a ritual my grandmother prepared three times a year without exception. In the center of our old and discolored Teflon cooking pan, she would burn the Palo Santo, along with apples, hissing oranges and limes, and cinnamon sticks and cloves that occasionally sparked in the heat, letting the smoke fill every room in our home. Before she began, she would scrutinize the rectangular pieces of bark with a discerning eye, sniffing to gauge the age of the tree and the natural scent and stickiness of the resin—key indicators of the tree's medicinal and spiritual potency. Following the *sahumerio*, we would bathe ourselves with roses and other flowers we'd hand-selected from the street market. I distinctly remember the steamy bathroom awakening all my senses. My grandmother and I seemed to share a genetic affinity for the allure of roses. This final step, after dispelling evil entities, invited in the spirit of tranquility, beauty, and hope that was living within the flowers. It didn't hurt that my skin smelled like roses for days after.

The traditions of *sahumerio* extend much further back than my grandmother's and Marina's era. When Spanish explorers first arrived in the New World, specifically Mexico in 1529, they documented the indigenous practice of burning copal, a tree resin closely related to the Amazonian Palo Santo. This was used along with other aromatic herbs for spiritual and medicinal cleansing. The Maya would add dried flowers or floral honey into their *sahumerio* rituals as offerings to the eagle goddess Chantico, seeking to attract love and romance. Sometimes, their *sahumerios* would burn day and night, which I understood to mirror the passion of a lover.

In Panama, archaeological charcoal from burned aromatic woods, typical in *sahumerios*, has been found in multiple burial grounds dating to around AD 1000. This suggests that the pre-Hispanic societies of the Isthmo-Colombia Area conducted *sahumerio* rituals during burials, probably to purify the souls of the deceased on their journey to the afterlife.

Marina's pot of water had reached a rolling boil. Tiny bubbles at the bottom of the pot grew from gentle whispers to louder murmurs, erupting on the surface. Each bubble burst and popped as the water simmered vigorously. Marina began to add the mountain rocks into the pot, one at a time. With each addition, the water hissed and sizzled, surging and occasionally splashing over the rim of the pot and into the crackling fire below. Through expressive hand gestures, she explained that the rocks would remain in the boiling water until they, too, were scalding hot. The leaves, twigs, roots, and Palo Santo would be used later. It was a different experience to the *sahumerio* I had grown up with, but I was fascinated by the intricacy of her process as she continued some of the oldest ancestral practices for traditional medicine that our planet has to offer.

Marina's niece came running toward her, pulling the bottom of her *cushma*, just as I would have done to my grandmother when she showed me the natural pharmacy she had built in our home garden. The young girl wanted Marina to see something she had found next to a large, incredibly thick tree—this, she articulated by stretching out her arms as widely as she could. Marina laughed with the same infectious energy that my grandmother had always shown me. At the age of six, when I dreamed of being an astronaut, I once dragged her out the warmth of her bed to gaze at *Las Tres Marías* (the Three Marys), part of the constellation Orion. When I was ten, and had moved on to a dream of becoming a nature doctor, I dissected *savila* (aloe vera) on our kitchen table, fascinated by its thick leaves and the gooey insides.

Marina's eyes reassured me that she would teach me everything I yearned to learn, and then she turned to exit the tent, matching her niece's pace. Before I showed myself out, I stood by the boiling water as it extracted the chemical potency of the plants. Thousands of molecules combined and joined forces in the pot. Similar to the art of making good tea, the water must boil to unlock the leaves' benefits. I wondered whether the special mountain rocks, or the potential microbes within them, had any biological impact on this process. Did the vapors carry volatile components with medicinal properties, acting immediately upon inhalation? I vowed to return to continue learning from the Ashaninka wisdom.

Back at the community center, Pascual announced it was time for him to help prepare the *masato* drink, or else Micaela would be seriously mad. I laughed as I tried to picture the wise *Apu* involved in a domestic scenario with his life partner, like those I had with my husband if he was running late with the milk for breakfast.

I have always found *masato* fascinating, as it embodies the transmission of knowledge through generations in the Amazon and demonstrates a sustainable use of land that brings joy to its people. Although you can now find it in fine restaurants that carry an Amazonian theme, *masato* is traditionally enjoyed from bowls crafted from dried fruits.

The Amazonian drink is made from fermented yucca. The process begins with peeling the starchy root, then grating and crushing it to extract the juice. In a time-honored tradition, Amazonian women may chew the yucca to begin the process of fermentation, adding natural enzymes from their saliva.[2] Mildly sour and effervescent, this drink is a cornerstone of Amazonian culture. It plays a vital role in social events, festivals, and rituals, symbolizing the Ashaninka's hospitality and community spirit. Beyond its cultural significance, *masato* is also central to how they relax and socially interact between arduous tasks.

There are thousands of other cultural and medicinal traditions that are rooted in the natural resources of the Amazon: plants, bark, leaves, animals, rocks, waters, soils . . . many life forms and ecosystems encoding unique chemistry and genetics that could contribute to regenerating our rainforests. Perhaps it is within the subtlety of exploration and indigenous wisdom that we will find the future of sustainability.

– *It's for the* kametsa asaiki *(good living)* – Pascual said with a cheerful wave, heading back to his home. As the clouds parted, giving way to a new range of mountains behind us, he added – *To living beautifully*, doctora.

2

A RIVER THAT BOILS

Yacu wasca

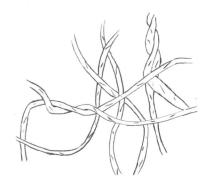

SCIENTIFIC NAME: *Doliocarpus* spp.

TRADITIONAL NAME: Yacu wasca

ORIGIN: Native to the humid rainforests of Central and South America

TRADITIONAL USES: Yacu wasca, where "yacu" means water in Quechua, is a climbing liana commonly found throughout the Amazon. It is known for its ability to store water in its roots and stem. When cut, the vine releases pure, crystal-clear water, accumulated from rain and humidity. It's like turning on a faucet in the heart of the jungle, releasing water with a subtle sweetness and a refreshing crispness. Amazonian people, with their profound ecological knowledge, can distinguish between the water vine and a similar but poisonous vine. The water from the latter has a bitter taste and causes short-lived intense diarrhea and stomachache, though it does not pose an immediate life-threatening risk.

SCIENTIFIC INFORMATION: Most *Doliocarpus* spp. are lianas, with many having developed an intrinsic system for water accumulation, allowing them to thrive in the challenging conditions of the Amazon. Although the liana can grow extensively, climbing high into the canopy to access the sunlight, its growth metrics have not been thoroughly documented. Thought to possess a unique blend of phytochemicals, little scientific investigation has been conducted to determine the chemical and biological profile of this lifesaving liana, for times when potable water cannot be found.

– *A river that boils?*

The Peruvian researcher with whom I had been conversing for the last half-hour scoffed from behind his wide wooden desk. Although we were in peak summer, the office felt 10 degrees colder inside.

– *We have no volcanoes in the Amazon, Miss Rosa.*

I reached for my bag, intending to show him the evidence I had meticulously gathered, including a copy of a TED talk I had recently viewed. In it, the scientist Andrés Ruzo revealed to the Western world the existence of a boiling river – a river so intensely hot that it literally boils, located deep within the Amazon. However, before I could get my hands on the folder I had so carefully prepared, he interjected, deliberately enunciating each word as if I was struggling to understand him.

– *Miss.* Here, he paused. *A "boiling river" would need a source of constant heat; it's simple.*

He stood up from his leather chair with his hands in his pockets, indicating it was time for me to leave.

– *Didn't you say you were a scientist?*

Some natural wonders have to be seen to be believed. Not only do they express ultimate beauty—they also harbor unfathomable mysteries that have eluded full scientific explanation. In the rainforest, for centuries, our ancestors accessed these wonders, attributing their formation and protection to powerful spirits.

I always knew I'd return to the forest. And it was right in the middle of completing my doctorate in Chemical Biology, and bouncing back from dozens of rejections, that I reached an incredible goal. This milestone was, in fact, the main reason I pursued a Ph.D. in the first place. After hours of time and effort, I obtained funding to come back home and study the lands of my heritage. This expedition proudly earned me the title of National Geographic Explorer.

I traveled over 5,600 kilometers (3,500 miles) from the Northern United States to Lima, the capital of Perú, to join an experienced and interdisciplinary team of explorers who were researching one of the most extraordinary and unexplained ecosystems I had ever seen in the lush rainforest. From Lima, we boarded a small plane bound for the Central Amazon, where we loaded our research gear and tools into 4x4 trucks and embarked on an off-road adventure across serpentine terrains to a secluded area.

After an hour or two on the rugged roads, our truck veered right, navigating through a maze of bushes, trees, and uneven hills. The wheels churned against the ground, kicking up clouds of sand that hung in the air like mist. The vehicle zigzagged: right, then left, straight, and left again. Clutching my equipment—and my stomach—I braced against the vehicle's G-force, silently promising myself that I would eat a lighter breakfast next time.

As we continued our descent, the road became less treacherous and the trees ahead grew taller. We were close. Months of intense preparation and years of longing had brought my scientific knowledge back home. It wasn't just the winding road that twisted my stomach. It was also the anticipation of reuniting with the ancestry of the rainforest after years of working inside a white-walled laboratory.

Between the vibrant greenery stretching toward the deep blue sky, the vista opened up to reveal endless steam columns rising among the tree canopies. If I didn't know any better, I would have thought that the steam was the result of a forest fire. Instead, here, the steam created an aura of wonder that seemed to pause time. If there were no volcanoes nearby, what could possibly be creating this much heat in the middle of the jungle?

The vapor rising from the scorching hot waters emerged with remarkable strength and volume, making it impossible to tell where the steam ended and the clouds began: a seamless union between land and sky. Even before it came into full view, the Boiling River announced its grandeur.

We exited the trucks carrying our luggage and scientific equipment, eager for some stable ground. As we approached our huts, I noticed a sign marked *"Agua Caliente"* (hot water) that was dripping with condensation from the steam trapped between the trees. I carefully approached the riverbank and was met by waters simmering so fervently that I could hear the powerful bubbles popping continuously and methodically. Blup, blup, blup.

The scorching vapor wafted toward us, and the heat and humidity stiffened our breathing, making each inhale heavy and thick. These vapors are thought to possess healing properties for both body and spirit, similar to the first Finnish saunas, created over 2,000 years ago, where water was splashed onto heated stones to generate steam for spiritual and physical purification. But in this corner of the Amazon, nature takes care of that on her own.

The Boiling River of the Peruvian Amazon is a remarkable ecosystem that continues to push the boundaries of science. It's an environment so extreme that over centuries it became the stuff of legends, with only the most powerful shamans daring to go near it.

A RIVER THAT BOILS

It is considered one of the largest documented thermal rivers in the world, flowing hot for nearly 6.5 kilometers (4 miles), reaching nearly 99°C (210°F) in some areas. The temperature is so hot that any small mammal that falls into it will instantly boil alive.

Unlike many other known hot springs, like those in Yellowstone, or remote regions of Iceland and Japan, the Boiling River is a non-volcanic thermal spring. Studies are ongoing to determine how the river boils. The most widely accepted theory is that the waters are heated to extreme temperatures deep within the Earth, flowing at great speeds along naturally occurring cracks that pierce through rock and sediment, finally reaching the surface and giving birth to a river that is constantly boiling. In line with the earliest written records, it's thought that the river's water has boiled for nearly two centuries, dating back to the era of the ancestors of one of the leading shamans in the area.

Through generations, the Boiling River has been revered as sacred. Its traditional name, *Shanay Timpishka*, roughly translates to "boiled with the heat of the Sun," and various points along the river are believed to harbor their own unique and powerful spirit.

If you look closely, somewhere past some of the hottest sections of the river, you will find a boulder that sits on the bank and is shaped like a giant serpent's head. There lies the *Yacumama*. "Yacu" means water and "mama" means mother, a mythical creature depicted as a giant snake that is the origin of the river's hot and cold waters, and continues to safeguard it against any harm. The *Yacumama* is considered the mother of all life in the water, maintaining balance in the ecosystem. Many believe that this spirit can cause a whirlpool and suck up any creature that dares to come too close, or may even spout water, causing flooding.

Another powerful spirit that lives within the river is the *Yacuruna*, meaning "the men of the water." Legend holds that the *Yacuruna* is capable of transforming into attractive and charming man-like beings who live in upside-down underwater cities, mirroring the human cities on land. These mythical Amazonian figures, often depicted riding crocodiles as canoes and using snakes as hammocks, are thought to lure the unwary if they act unwisely, abducting humans and bringing them to the depths of their underwater homes. The *Yacumama* and *Yacuruna* spirits are both considered the guardians of the river, faithfully protecting one of the most enigmatic corners of our planet and inviting us to embrace the spirituality of nature and the wisdom of our ancestors.

During our expedition, the Boiling River revealed intriguing secrets and taught us much, in both the realms of scientific curiosity and spiritual insight. One of the deepest lessons we learned was to be prepared for the unexpected, a lesson that would forever impact my view of human resilience and teamwork in times of risk.

Early in our travels, the team and I decided to hike to one of the most extreme sections of the river, among a space of shifting geothermal grounds. The terrain felt spongy and bouncy, and uncertain, like shifting sand. This prompted us to use walking sticks as we traveled, tapping the ground before each step to secure our footing.

Tap, tap, tap. Step. Tap, tap, tap.

We had to cross over the fierce, bubbling waters of the river, stepping on perilous logs and boulders as we ventured deeper into uncharted territories. Each step required meticulous calculation. With my left foot secured on a small ledge in the rocks, I carefully assessed the next safe landing spot for my right foot. Balancing forward over the rumbling river below, we all clung tightly to our walking sticks, which braced against the gurgling, popping waters. The slippery algae added a mischievous challenge as we fought against the growing current. Crossing the Boiling River demanded extreme focus and attention. There was no room for mistakes.

After a demanding journey, both physically and mentally, we finally arrived.

We spent hours exploring the surreal landscape of spongy terrains until most of the group headed back to camp to feast on a well-deserved lunch of rice, beans, and bananas. Meanwhile, seven of us stayed behind to conduct a few additional experiments: Andrés, the geo-environmental scientist who'd first studied the river and was a soul brother to us all; Emma, a geothermal expert and wise woman; Leo, a field biologist and poet; Lucas, a multimedia storyteller and naturalist; Joao, a biologist and conservationist; and Marshall, a clinical mental health counselor and outdoorsman.

We shared stories and jokes as we recorded scientific data. It felt like the best kind of science camp, but one where our work could have meaningful implications for the river and rainforest. Before we knew it, the sun began to set. Without the right gear for night exploration, we decided to wrap up our work, pick up our tools and start heading back to camp.

But just after we stepped onto the hiking path, everything changed. We heard a loud, intense scream that pierced through the rainforest. My throat immediately closed as I looked around, terrified. My mind raced to find logical explanations, but nothing could have prepared me for what we saw next; Marshall had stepped into boiling mud.

Rather than twigs and leaf matter crunching below his feet, his foot had sunk ankle-deep into thick, scorching mud that exceeded 80°C (176°F). He wore only sandals, and within seconds, he'd suffered severe third-degree burns. My heart sank to my stomach. Marshall was the tallest among us, so even if we tried, we wouldn't be able to carry him without risking further injury. But we needed to reach camp fast.

We started hiking out of the heart of the Amazon, determined to find help, with Marshall painfully but resolutely walking alongside us. The sun slowly dipped below the horizon, the skies orange and pink as we moved.

– *It's only thirty minutes down this path* – we said in quick succession, and we started singing songs and recounting jokes to keep Marshall in good spirits.

The route was still freshly trodden from the morning. Lianas intertwined at every step. The trees and bushes around us were so tightly packed together that we needed to walk in single file. And then the fire ants arrived.[1] The ants stung Marshall repeatedly through his sandals, causing intense burns and exacerbating Marshall's pain. Then came moist, cold, and malleable soils into which our feet sank with every step. The songs stopped, and darkness closed in.

None of us, a group of experienced explorers of the Amazon, had prepared for staying out past daylight. With one headlamp and four phones in hand, we shined a light two feet in front of us, carefully trekking down the muddy hill. We could hear Marshall cry from the agonizing pain. The burns could cause irreparable damage to his nerves and skin, which might lead to limb dysfunction and require extensive reconstructive surgery. Every minute counted. He needed a nurse *quickly*.

That's when Andrés, who was the most experienced among us at navigating this terrain, made the valiant decision to run ahead, through the jungle, and head toward the village to look for help. The rest of us kept trekking until we saw a faint light that indicated we were approaching open ground. A cheer broke out, but it was quickly silenced. As we exited the path, we had one more challenge to face. The Boiling River looped back round the forest to block our exit. Only one perilous log and a few boulders crossing the river stood between us and our safe return. My heart pounded, and everyone was suspended in a state of uncertainty. In the expansive darkness, all we could see in front of us was the scorching hot vapor rising from the river. Bubbling, hissing, rumbling. It was so dark that the vapor blurred reality, making it seem like we were trapped in a lucid nightmare.

A gust of wind swept through, temporarily clearing the thick haze of steam, presenting in detail the single log that bobbed at the mercy of the turbulent waters. Precarious and unsteady, one miscalculated move and we would all fall into these ebullient waters. In trying to save someone's limb, we would risk everyone's lives. That was the moment we realized we were stuck. We weren't going anywhere.

Surrounded by dense and uncharted rainforest, with impassable boiling waters ahead, a shift occurred within our team. Our sense of exploration transformed into an ancestral instinct for survival. The rainforest demanded our attention. Within seconds of realizing we were stranded until further help could reach us, we naturally fell into roles dictated by our instincts. After gently helping Marshall to lie down, Emma and Lucas began illuminating our surroundings, deterring ants, tarantulas, and any creeping critters that seemed drawn to our plight; Leo used his headlamp to send signals through the vapor cloud, marking our position; Joao and I crouched beside Marshall, with Joao aiding in cooling Marshall's injury by fanning and dripping the last of our water over his feet, while I frantically searched my backpack. Buried beneath my sample-collection kit, notebooks, and snacks, I found my first-aid kit. Instinctively, as if guided by my grandmother's words, I seized the thermal blanket. Marshall had started to shiver, and we feared him going into shock. I couldn't let that happen. I wrapped the blanket round his body and hugged him to preserve heat.

– *What is your favorite food, Marshall?* I asked, hoping to distract him, my voice tight with emotion. *What is the first thing you will eat when you are back home?*

– *My mom... she makes this... it's delicious... could eat... she cooks... so well...* Though his speech was halting and strained, and his sentences incomplete, Marshall's heartbeat and breathing began to slow, and his tremors to ease, as he recounted stories of his mother.

– *I'm not mad at the river* – he whispered, clutching my hand, his lips curving into a smile. *I just walked where I shouldn't.*

The rainforest has a way of teaching us our place in nature like nowhere else. The communities in the forest deeply understand that the Pachamama, the goddess of reciprocity and duality, demands respect. In the Andean-Amazonian cosmovision (see page 16), people see nature as a living entity, deserving of protection, respect, and nourishment—so that people, in turn, are protected, respected, and nourished. It's a cycle of giving and receiving, just as the Incas held grand ceremonies to honor the sun and request prosperous crops, or the Amazonians pray to the *Avireri* mountains before hunting or gathering, promising to only take what they need and to leave enough for those who come next. Just as they did for the Incas, these rituals and practices allow Amazonians today to continually express their gratitude to the natural world for providing them with food, medicine, shelter, and life. They understand the immense generosity of the Pachamama and the importance of respecting her. If these acts of reverence are neglected, they fear the wrath of Pachamama in the form of floods, droughts, or pests, as a reminder of the balance they have been entrusted to maintain.

As Marshall lay on the ground, the ethereal white-gray steam that enveloped us suddenly lit up. The beams of tens of cellphone flashlights shone on us, as if we were at a music concert with smoke machines and spotlights. Word had reached the rest of our group. Many stood along the riverbank, while others, braving the currents, had leaped across the river's largest rocks.

Upon hearing of Marshall's accident, the local community had mobilized to gather wood and construct a bridge as quickly as they could to assist us. The Shipibo-Conibo live in the Peruvian Amazon as an indigenous community, known for their artistic expressions of the Pachamama. They know not to defy the temperamental river. But equally, they know that the spirits residing within these waters are as benevolent as they are powerful.

What unfolded over the course of an hour seemed to flash before my eyes. Andrés had returned, and three community members followed, carrying wood, working. The Shipibo-Conibo men skillfully maneuvered a lengthy wood panel, positioning it so it generously overlapped the boulder we stood on, providing a stable platform. Panel by panel, they constructed a sturdy bridge that could confidently hold two people at once. As the wind strengthened the river's current, the wood panels quivered, but the men hammered long nails through, securing the path across the bubbling waters. One of them even tested the bridge by bending his knees and briefly jumping on it. I gasped. The ultimate testament to Amazonian bravery.

The men crossed the new bridge, one carefully lifting Marshall to transport him across, with the other two on either end wielding sticks for us to hold for added stability as we crossed the river's capricious currents. With the mastery of someone who has long known this river, the Shipibo-Conibo man confidently transported Marshall to safety. Everyone began clapping and cheering with big smiles, fists in the air, and sighs of relief. We were safe.

Upon reaching the other side, a nurse quickly assessed Marshall in a makeshift recovery area. Despite a long healing journey ahead, he wouldn't need surgery. But it was a night that none of us would forget, and a valuable lesson from the rainforest.

For the rest of the trip, I was affectionately called "Mama Rosa," due to the maternal instinct that had taken over me when I was caring for Marshall. Curiously, no one knew that my grandmother was also named Mama Rosa. As a young girl growing up in the high-altitude mountains, her translucent white skin would bloom with a red glow on her cheeks, making her the most beautiful girl around. But she also happened to be one of the most devoted people in her town, always finding natural cures within her garden and spirit to ease people's pain and discomfort. These traits earned her the eternal nickname of "Mama Rosa," and I was quietly honored to be sharing it. I have no doubt it was her spirit that guided me to care for Marshall that night.

The Boiling River is an extreme within the extreme—a river that boils in the heart of a tropical jungle. We had come to this river to uncover life forms that were uniquely adapted to thrive in seemingly inhospitable conditions, surviving where nothing else could. For years, the existence of such a natural wonder in the Peruvian Amazon was met with skepticism from scientists, researchers, and government officials alike, who laughed at anyone who even dared to pose the question. I'd seen that happen myself. But new discoveries come about by observing and listening with childlike wonder, free of presumptions; by embracing the vastness of what we have yet to learn; by acknowledging that, although the human mind is extraordinary, the Pachamama is more extraordinary still.

Amid the fierce conditions of the Boiling River, a tapestry of microscopic thermophiles thrive in plain sight. Thermophiles are heat-loving organisms: *thermos*, meaning "heat" in Greek, and *philos*, meaning "loving." These microbes have a preference for high temperatures, typically 45–122°C (113–252°F),[2] and are found in hot springs, hydrothermal vents at the bottom of oceans, volcanic areas, and geysers. Their ability to thrive in the extreme can provide insights into the early evolution of life.

These ancient life forms, some firmly anchored to the stones and others free-floating in the waters, have developed their genetic and biochemical compositions over thousands of years to flourish amid the river's lethal conditions. They play a crucial role in sustaining ecological equilibrium, yet have eluded scientific inquiry until now in ecosystems that are central to our life on Earth, like the Amazon.

Many still ask how they can survive in such extreme heat. The answer is that thermophiles have specialized protein structures that provide them with the capacity to remain stable and functional at high temperatures. These proteins contain heat-resistant amino acids with unique folding patterns that prevent denaturation (damage), keeping DNA from melting. Their cell membranes are rich in heat-stable lipids that allow them to maintain integrity in boiling conditions, avoiding disintegration. Millions of these microorganisms are tirelessly cycling nutrients, enriching our soil, fortifying our forests, capturing carbon, and emitting oxygen.

The Boiling River acts as a natural laboratory for studying climate change. Its extreme environmental conditions create a microclimate so intense that it should be inhospitable to life. Yet, flora and fauna thrive around the bubbling waters and scorching vapor, demonstrating that in the depth of the Amazon, microbes have adapted to sustain life under these harsh conditions. The thermophiles found in the river's waters and rocks are not only capturing and storing carbon at high temperatures, but also activating dark parts of their genome to produce new molecules with significant implications for medicine, bioremediation, and agriculture. These genetic and biochemical adaptations could inform the development of sustainable solutions, offering heat tolerance to trees, plants, and crops in the face of rising global temperatures.

On another day, after an hour of trekking, we found a waterfall, forcefully descending into an immense pool of crystalline water filled with an array of vibrant aquatic life, thundering and dominating in this secluded paradise. The waters splashing over rocks seemed welcoming. If you didn't know that these waters were still dangerously hot, you'd be tempted to jump in for a refreshing swim. But you'd better not.

The waterfall pool immediately captivated us with its multitude of underwater life. Contrary to the rapids, which were dominated by boulders, this area revealed a submerged jungle deep beneath us.

In the pool's heart, we observed what seemed like a forest of miniature trees interspersed with warm springs. These two-inch-tall algae, numbering in the hundreds or even thousands, formed an underwater canopy. The tree-like algae undulated with the water's flow, dancing rhythmically, reminding me of the rainforest tree cover waving in the wind when observed from the sky.

Beneath this green layer, a vibrant array of microbial mats unfolded, resembling a multicolored cake. Such mats are often an indicator of healthy conditions in a lake. Layers of pink, orange, blue-green, and black microbes, each housing countless organisms, crafting an optimal ecosystem for their survival. These layers, each uniquely adapted to the prevailing conditions of heat, sunlight, nutrients, and waste, coexist in harmony, perfectly balancing one another in a friendly competition where everyone wins.

A flicker of movement caught my eye in the water. Closing my eyes for a moment to ensure my mind wasn't playing any tricks on me, I found the rest of our team had also noticed something out of the ordinary. Gathered around a small pool of shallow hot water, where temperatures likely hovered at 40–45°C (104–113°F), we spotted tiny red larval insects. They didn't seem to mind the heat, darting about in a large circle atop dense, black microbial mats anchored to the rocks below. Were these thermophilic insects? I had never seen anything like it before. How had they adapted to be here? Was there any symbiotic relationship between the microbial mats and the larvae to help the insects survive increased temperatures? We watched in silent awe, observing their chaotic dance as they intertwined and playfully clambered over one another.

Although rare, insects surviving in extreme conditions have been previously documented, with some species in desert environments, like Saharan ants and Death Valley beetles, or in thermal features, like ephydrid flies in Yellowstone, developing adaptations to withstand intense heat. Their body adaptations might include mechanisms for efficient heat dissipation, or they may produce heat-shock proteins that protect them against irreparable damage from extreme temperatures.

Under the microscope, before we gently returned them to their habitat, these tiny larval insects resembled a kind of beetle but with lighter orange hairs that may help reflect sunlight, reducing heat absorption and helping keep them cooler. This might have been the first time these thermophilic insects were noted in the Boiling River, which highlights how much undiscovered biodiversity in the Amazon is still left to be explored.

Fortunately, no more life-threatening accidents occurred during the rest of our time at the Boiling River. Instead, we constantly adapted our scientific methods and pace to match the river's rhythms. That's how science unfolds in the Amazon—and we learned that the hard way. Rather than following calendar reminders and phone alarms, we surrendered to the cycle of the sun and waters to guide our daily activities. It was a lesson in versatility and humility, reminding us that the will of nature is stronger than man-made timelines.

Embracing this process of surrender became a daily practice in the field, instilling in me a sense of freedom to rejoice in the beauty of the outdoors. One night, I found myself lying down next to the river, on a warm, giant-sized rock. The steam rose, coiled and faded in the midnight sky, creating a veil of ethereal charm against the towering trees that stood like silent guardians. Water streamed and splashed as if stubbornly fighting its way through ancient structures. The thunderous roar and high-pitched splatter of the current added a bass drumming and cymbal clinking to the uncountable sounds around us: dozens of crickets chirping; frogs rhythmically croaking and slapping their gentle feet against leaves and wet surfaces; insects flying and whirring, creating a fluttery wave of sound behind them; bats squeaking, their wings flapping a short distance away.

– *Did you see that?!* squealed Andrés with childlike excitement.

I had been distracted by a giant tarantula crawling between the rocks, but I looked up to the pitch-black sky to see not one, not two, but three shooting stars flying through the limitless Milky Way, who had revealed herself since the sun went down. Unconsciously, I gasped and held my breath, not wanting to interrupt what was unfolding. After a minute, and a slight rush of blood to my head, we counted our blessings, thinking we couldn't have had a better view. Suddenly, an explosion of color, akin to a firework, filled the sky. One star after another, spheres of light and plasma, flew over our heads, irrevocably and unintendingly carrying millions of people's wishes and dreams. A glittering stage for the dense star fields and rifts of our galaxy, with brilliant trails of light cutting through the darkness. Some of these meteors were only quick flashes, gone before I could even savor them, like neural connections that fire away in our brains when we recognize an old smell that lights up a memory from childhood. Others, to my delight, moved much slower, patiently leaving long, glowing trails behind that faded slowly and varied in brightness, allowing for my breath to return.

That evening marked the most mesmerizing starry night I had ever experienced in the darkest Amazon Rainforest. Land, water, and sky, all seemed to conspire, compelling me to be present in the now. I felt my fingers touch the uneven ground; my nose perceived the chemical complexity of the river, steam, and surroundings; my eyes filled with light from the sky and the moon, reflecting the wilderness of the forest. It was the most aware of my body I had been in a while, yet the most transcendent I had felt, as if my being extended into the natural wonders around me, absorbing strength and knowledge that came from deep within. Perhaps it was the perception of magic in the sky, the convergence of multisensory experiences, or the feeling that this was how my grandmother had always described the Pachamama: infinite, generous, accepting, all-giving. Like an everlasting hug that keeps on hugging.

Ultimately, our work is one of raising awareness—not only to protect the sacred Boiling River, but also to highlight that the Amazon's mega-diversity does not end with what the naked eye can see. Everywhere on our planet, life "hides" in plain sight, flourishing in the most extreme of environments, much like diamonds beneath the Earth's surface. Just as our teamwork infused us with strength, courage, and laughter to overcome the unimaginable, and just as discoveries come to light when we dare to be adventurous and open-minded, we have learned to surrender to the experience of exploration, and to embrace humility. The Pachamama has so much to reveal if we are willing to listen closely.

3

LIFE IN THE WATER

Camu-camu

SCIENTIFIC NAME: *Myrciaria dubia*

TRADITIONAL NAME: Camu-camu, caçari, araçá-d'água

ORIGIN: Native to the lowland floodplains of the Amazon Rainforest

TRADITIONAL USES: An Amazonian native fruit long valued for its nutritional and medicinal properties. This delicious fruit is consumed for its immune-boosting effects, attributed to its high vitamin-C content, considered one of the highest found in plants worldwide. Medicinally, camu-camu is used to treat a wide range of infections, enhance mood, support reproductive health, and improve or prevent degenerative eye diseases. It is also popularly used as an anti-inflammatory agent, alleviating colds and fortifying the nervous system. It's an essential component in local diets, and is mixed into smoothies and juices that may look bright pink, or be eaten fresh, providing an exquisite tart flavor similar to that of lemons and cherries. Now considered a "superfood," camu-camu is available as a powder, capsules, or extracts in markets worldwide.

SCIENTIFIC INFORMATION: Camu-camu grows as a bushy riverside tree up to 3–5 meters (9¾–16½ feet) tall, harboring small flowers with white petals, purple-red cherry-like fruits, and a sweet aroma. The fruit contains up to 60 times the amount of vitamin C found in oranges. When processed into juice, its anthocyanin molecules can react, creating a bright pink color. This fruit is also rich in antioxidants, flavonoids, and amino acids, including leucine and valine, which are critical for muscle health, combating inflammation and oxidative stress. Scientific research has revealed that camu-camu may play a role in reducing chronic disease markers related to obesity, cardiovascular health, and metabolic syndrome, though further research is needed. Furthermore, its vitamin C and ellagic acid content have been related to anti-aging properties and diabetes prevention. Camu-camu's unique chemical profile has earned it a position as a superfood in dietary supplements, health foods, and beauty products.

– *Something is pulling the fishing pole!* I shouted as I carefully braced my knees, trying to stand without losing control of my pole, which was being pulled taut by a large, native fish.

– *Hold tight!* exclaimed Rufilio, our fearless guide, as he quickly stood, his own pole clanking onto the boat's floor. He pushed aside the rest of his tools and fish bait.

I observed the murky brown waters below our boat, which were teeming with aquatic plants, shrubs and trees piercing the surface. We were in a semi-aquatic flooded ecosystem in the Northern Amazon, learning traditional fishing techniques to catch piranhas. An endless corridor of green and brown foliage melted into the water's depths in perfect, glassy symmetry. Some branches and leaves, partially submerged, moved softly with the wind, swaying in the river. I couldn't quite catch a glimpse of what was hiding below, except that it must have been of considerable size to add so much weight to my pole. All I could feel was a tense vibration along the line, my traditional wooden fishing rod bending into an arc.

A shadow of movement stirred the water, disrupting the reflections of the flooded forest. And before I knew it, the pulling stopped.

– *Ahh . . . so close* – Rufilio said, tapping me on the shoulder before leaning forward to observe the dozens of tiny bubbles rising to the surface. *You will catch one soon. Gotta catch 'em all!* he added, bursting into laughter at his own *Pokémon* reference.

I shook my head, making a mental log of how much of popular culture has made its way into the most remote places in the Amazon. I joined in with Rufilio's laughter, aware that if our survival depended on my fishing skills, we would go hungry. I began to put another bait on my fishing pole when a shout came from behind me.

– *I got it!* It was Chris, my husband, who often joins my expeditions. He had effortlessly caught the largest piranha we had seen so far that day, measuring about 20 centimeters (7 ¾ inches) in length. It had a pale, silvery hue with a gradient of colors darkening toward its dorsal side. Patches of an orange-reddish color, particularly around the gills, indicated it might be a red-bellied piranha (*Pygocentrus nattereri*). Its glossy skin was marked with small, dark spots covering its whole body, which appeared as tiny specks. The reflective scales caught the sunlight, providing a natural camouflage in its freshwater habitat.

Rufilio carefully held the piranha and gently opened its mouth to reveal its infamous teeth and jaws: a single row of sharp, triangular, interlocking teeth in both jaws, enabling the fish to slice through its prey. Piranhas can replace their teeth throughout their lives, and their powerful jaw muscles allow them to exert a force up to 30 times their own body weight, granting them one of the strongest bites among bony fish in the world.

If you thought piranhas were intimidating, consider their ancient ancestors, which lived in our waters 8–10 million years ago. Known as "megapiranhas," they were considerably larger, reaching up to 1 meter (3 feet) in length and about 11 kilograms (25 pounds) in weight. At least ten times larger than their modern descendants, their jaws delivered a bite force that was considerably more powerful by at least an order of magnitude. Experiments with a replica of the megapiranha's teeth have shown that this creature's bite could have penetrated the shell of a turtle or even bones. You're welcome.

– *Piranhas' teeth make for great traditional sharpeners of wooden tools, like darts* – Rufilio told me, mimicking a back-and-forth movement as he slowly brought the piranha closer to me before placing it back in a bucket filled with other, smaller fish for dinner later that night. Without any scientific query or purpose in mind, driven purely by the curiosity of a child, I had taken my non-invasive field microscope out of my bag. What does piranha skin look like when magnified?

I've learned that when we have the ability to look at nature up close, we unlock an array of colors and textures hidden from the naked eye. It's like accessing new palettes of color that satisfy the right part of my brain. Curiously, the fish's orange patches dotted with freckles looked like the veins of a silky flower, extending outwards in a symmetrically pleasing structure. The rest of its silver body reminded me of snake scales when seen up close. Although fish and snake scales have many physiobiological differences, they both contain keratin, the same natural protein found in human hair and nails.

Despite their undeserved reputation as bloodthirsty and ferocious predators, piranhas eat insects, crustaceans, fish, and plant material, and some species have even been observed consuming seeds and fruits that naturally fall into the water. They travel in large groups known as "schools" to protect individual members from potential predators like caimans (Amazonian reptiles closely related to alligators and crocodiles), birds, and dolphins.

The 80s movie *Piranha* gave these fish a bad reputation, depicting abnormal, exaggerated behavior. The film was inspired by a rather unique incident involving former US President Theodore Roosevelt. During a 1913 expedition to the Amazon, locals in Brazil were looking to impress the president, and so they starved a group of piranhas and then placed a live cow in a netted-off section of the river, inducing a feeding frenzy that was described as showing "evil ferocity." This act captured the world's imagination, convincing many that piranhas were more aggressive than sharks and a life threat to humans, fueling movie scripts and cementing their fearful reputation.

Many people in the Amazon, including locals and professional open-river swimmers, regularly swim or train in piranha-filled waters without coming to any danger or harm. These native fish won't typically nip unless hurt or attacked—there are stories of distracted fishermen who have lost some fingers.

Unintentionally, when I was 12, I too swam with piranhas. My mom and I were in the Amazon visiting some distant family when we decided to join some local tours up north. We went piranha-fishing with a local community when one small piranha suddenly jumped into our small *peke peke*,* landing right in my mom's lap. After a high-pitched scream of surprise, she started laughing loudly, calling it a sign of "good luck" as dinner had quite literally landed on her lap. After a few hours, we headed to a small mud-filled island in the middle of one of the Amazon's tributaries. We dipped into the refreshing waters along with other families who were enjoying the sun and time off school. I remember seeing glassy reflections of various-colored fish swimming around us, including small orange and silver-tinted piranhas moving right and left near us. At the time, I had never heard of the popular myths or watched the multiple movie remakes in which these fish are depicted as ferocious predators. To me, they were simply another life form in the Amazon, brimming with life and flashing with color.

*

---

* A *peke peke*, my all-time favorite ride, is a traditional wooden Amazonian boat, a modest yet skillfully constructed vessel powered by an electric motor at the rear. Its name derives from the sound the motor makes in the thick and dense waters of the Amazon. I suggest you try saying the name out loud very fast, many times, to get a pretty good idea of how these boats sound in the field.

After storing all the fishing gear, we continued our journey down the slow-moving tributary looking for food, whether fish or fruits, in the waters of the Amazon. In one swift motion, a kingfisher with vibrant plumage appeared out of nowhere. Its body was streamlined, with a large head and long, pointed beak aiming for prey with remarkable speed and precision. The bird, which had a white chest and a hint of orange around its neck, plunged into the water and re-emerged with a small fish caught in its beak, still flipping its tail, before it flew away, disappearing out of sight.

Nearby, a large bent tree on the riverbank extended its long sticks and branches into the water, creating a small, shallow pool where a Cocoi heron (*Ardea cocoi*) stood elegantly. It perched atop a small rock on its slender, long legs, its body white with gray feathers and small black patches, its eyes piercing the surface of the water. It seemed like everyone was looking for food along the tributaries of the Amazon River.

We advanced along the riverbank, approaching a space where various 2- and 3-meter-tall (6½–10 feet) trees were partially submerged in the water. They had evergreen leaves, and thin trunks and branches. Plant matter accumulated on the surface, giving life to a unique microecosystem. I could see dozens of bubbles constantly coming up to the surface from plants that had made their home underwater, indicating continuous oxygen production.

Hidden behind the lush treetops was a local Amazonian man aboard his one-man *peke peke*. Dressed in shorts, a gray T-shirt, and a cap, he had a blue plastic case behind him. When he noticed our presence, he waved happily before resuming his task.

He was collecting camu-camu fruits (*Myrciaria dubia*), one by one, branch by branch. Each tree harbored hundreds of small green-red round fruits that resembled cherries or grapes from afar. As he grabbed each fruit, he quickly deposited it into his case, which by now was half-full. Whenever the slow and soft movement of the water edged him further from the tree, he would push himself forward with his paddle.

I immediately thought of the delicious camu-camu ice cream I'd had a few days before. In fact, I had just had camu-camu juice for breakfast. I like to call it the Amazonian pink soda. This native fruit is packed with vitamin C, and its nutritional and medicinal properties have been widely recognized, fueling its commercialization; it's soon to become the next açaí.

The camu-camu fruit is sometimes used as bait for fishing. Some species of native Amazonian fish are attracted to it, as the ripened fruit can be found floating in the water. The fruit's skin is repurposed as a natural dye for fabrics, and the tree bark, roots, and leaves are also essential components for licorices and traditional medicinal syrups. Curiously, it's also a preferred food for Amazonian native bees, which contribute to its pollination, boosting its crop production by up to 44 per cent. Some of the native bee honey can even taste distinctively like camu-camu, which thrives in lowland, swampy, floodplain areas influenced by the seasonal flooding of the Amazon River and its tributaries.

As we steered forward, the slow-moving waters started to change, and we heard a gentle but continuous murmur growing. We were approaching a small rapid created by natural rock formations. That was our sign to navigate away from the riverbank and into the center of the tributary, veering right at the next turn, as we rejoined the majestic Marañón River.

*

The sun was shining high in the sky, the water reflecting an infinite spectrum of colors, creating shadows and textures that served as a guided meditation for my brain. I was surrounded by the constant sound of water moving and flowing, connecting land, people, plants, and animals with the unknown depths of what lies at the bottom of the Amazon River and its tributaries.

White foam formed next to the boat as the motor's propellers stirred up the water and air, agitating organic materials and sediments. The white contrasted with the tea-like color of the Amazonian waters. This characteristic brown color arises from tannins—natural molecules released from the decaying plant matter of trees, shrubs, and other local vegetation—dissolving into the water.

I settled in, resting my feet on the boat's bench, and was about to close my eyes for a minute when I spotted a vanishing pink shadow in the distance. I stayed still, carefully dissecting the space. And just as I thought it might have been a figment of my imagination, the water whooshed with a rushing noise.

A dolphin broke the surface, followed by a loud splash as it leaped back into the water. And in a cascade of beauty and mysticism, against the evergreen backdrop of the Amazon, a second, third, and fourth dolphin followed, as everyone gasped in admiration.

– Bufeo colorado! *A whole family of them!* proclaimed Rufilio with a big smile, extending his fist in the air with excitement.

These weren't just *any* dolphins; they were Amazonian pink dolphins.

Also known as *bufeo colorado* or *boto*, this unique kind of freshwater dolphin species is found in the Amazon River. Known for their distinctive pink color, with some even achieving a full flamingo pink, they are born gray and gradually turn pink as they age, with males becoming drastically pinker to attract more attention from females. Their coloration is considered a result of scar tissue from constant abrasion of their skin, either from fighting or playing rough. However, their final color is influenced by diet, behavior, exposure to sunlight, and other factors.

Unlike other dolphins, their flexible necks have unfused vertebrae, allowing them to turn their necks 90 degrees and maneuver around rocks and tree trunks, while their echolocation enables them to navigate the muddy waters of the river with ease. They don't call them the gymnasts of the Amazon for nothing.

– *The legend says that the* bufeo *can transform into handsome men by night* – said Rufilio, sitting with his elbows resting on his knees. He turned his cap backward, his long black eyelashes extending as he opened his eyes wider, and immersed himself in the storytelling.

– *Some even say the* bufeo *may dress up in a white suit and a white hat, the hat covering the blowhole still atop their head, as their shapeshifting abilities aren't always perfect.* He pinched two corners of his shirt with his fingers, moving it back and forth, mimicking a gentleman stroking the fabric of an elegant suit, laughing.

– *And they seduce women . . .* I said in a low voice, always loving hearing Amazonian myths from different voices and people.

– *Yes! People say the* bufeo *is so seductive that they can make women pregnant just by seducing them and staying close to them* – Rufilio agreed, rolling his eyes upwards.

These pink dolphins are closely associated with *Yacuruna* (see page 35), legendary aquatic beings that resemble humans and live in underwater cities. Considered powerful healers and masters of the waterways, *Yacuruna* are thought to have god-like powers that enable them to communicate with aquatic animals in the Amazon, like the *bufeo*, and to abduct humans and take them to the bottom of their underwater homes.

Pink dolphins have a reputation for guiding fishermen to areas rich in fish, yet some also claim that the *bufeo colorado* may lead navigators into peril, luring them into treacherous waters and sinking their boats. They have also been observed throwing sticks and weeds and grabbing fishermen's paddles with their beaks. In many Amazonian cultures, eating and harming pink dolphins is considered bad luck.

– I think they may be the smartest animal in our waters – whispered Rufilio.

Amazonian pink dolphins are known for their unusually large brains, with research suggesting self-awareness in dolphins through self-recognition in mirrors, an attribute previously thought exclusive to humans and great apes. Dolphins have been observed mourning their dead, carrying deceased individuals on their backs for hours or days, showing empathy by assisting the ill, teaching skills to one another, and living in complex social groups.

It is the varied range of responses and neurological skills observed in dolphins, as well as in other endangered Amazonian native species, including bees, that has inspired our team of scientists, conservationists and environmental law partners to advocate for the rights of Amazonian animals to exist as sentient beings, a movement known as the Rights of Nature. This movement recognizes the rights of life forms and ecosystems to exist and regenerate, granting nature the right to be represented in court and protected.

In early 2024, in an incredible and unexpected turn of events, our partners, the Earth Law Center organization, along with an incredible local community led by indigenous Kukama women, succeeded in having the Marañón River recognized as a subject of rights with intrinsic value,[1] with the indigenous communities designated as its guardians and representatives. This marked a revolutionary step in conservation in Perú.

The two smaller pink dolphins leaped out of the water again, emitting a unique whistle and a buzzing sound, like a sequence of pulses, followed by playful squeaking as they splashed in the water, creating ripples around them. Swimming in front, the two other pink dolphins remained covered, with the tips of their fins occasionally revealing their presence.

I looked around as we all smiled, mesmerized by the beauty of the mystical *bufeo colorado*. It felt as if we were witnessing a unique ritual of animal play.

Rufilio cupped his hands around his mouth, forming a concave shape with his palms and fingers, creating an improvised sound tunnel. He started emitting what felt like a textured whistle intertwined with sharp clicks. It sounded remarkably similar to the dolphins' calls.

– *Just saying hi to my friends.* He smiled, the lights in his eyes dimming. *It's getting harder to see them nowadays.*

Amazonian pink dolphins are considered the largest freshwater dolphins on our planet, reaching lengths of up to 2.5 meters (8¼ feet), and they consume more than 40 kinds of species of crabs, turtles, shrimps, and particularly fish, including the infamous piranha.

Their secretive nature has contributed to the mysticism surrounding their existence, making it difficult to track their paths or fully assess their distribution, as they tend to be shy and elusive. This behavior, along with the myths associated with the species, may be what has allowed them to continue to exist, even as their populations and ecosystems are increasingly and dangerously threatened by mining contamination, dam construction, deforestation, pollution from agricultural runoff, and other risks.

Studies revealed in early 2024 indicated that their ancestors lived in the Amazon more than 16 million years ago and grew up to 3.5 meters (11½ feet), making them the largest river dolphin ever known to science. The leading scientist found part of an ancient dolphin's skull sitting on a riverbank while walking across the Amazon. Over the years, fossils of many giants have been discovered in the Amazon, but this marked the first dolphin of its kind.

A friend of mine once told me she saw a family of Amazonian pink dolphins swimming and playing in a deeply flooded section of the river, with tall trees submerged as the dolphins leaped among the treetops. Pink dolphins swimming between tree canopies—a mesmerizing image I hope to see myself one day.

Their procreation season occurs during the rainy season, allowing males to leave the flooded rivers and swim for miles between the flooded forest trees. This enables them to search for prey and fish while also reaching females that reside in lakes, caring for their calves. As the dry season approaches, the males return to the Amazon River and its tributaries.

As we continued our journey along the Marañón River, a meandering river in the lowland floodplains of the Amazon, we eventually approached the point of convergence with the Ucayali River. We had moved into an expansive body of water that felt infinite, save for the thin line of distant Amazonian canopy and trees that bordered both sides, creating a sense of embracement. Reflections glistened in every corner, and a few large clouds decorated the blue sky.

– *That's where the Amazonas begin!* Rufilio exclaimed, pointing at the center of the never-ending waterway. He was indicating the confluence of the Marañón and the Ucayali Rivers, the point that marks the birth of the Amazon River, also known as the Amazonas in Perú. As the two tributaries feed into the Amazon River, they provide the largest volume of flowing water compared to other tributaries.

The birthplace of the Amazon River, where two long and powerful rivers merge, is considered a sacred and lucky spot, with many making a special wish upon arrival. I noticed Rufilio murmuring some words to himself.

I made my own wish, then sat in silence for a few minutes, observing the infinite landscape, marveling at the ebb and flow of water throughout the Amazon.

All native life forms, including Amazonian pink dolphins, constantly adapt to the various water-based ecosystems that emerge, then disappear, only to return once more, following the perennial water cycles. It is this complexity that breathes life into one of the most biodiverse regions in the world, where human, flora, fauna, fungal, and microbial life has evolutionarily adapted to thrive and flourish in these changing conditions.

I looked to the east, imagining the Amazon River stretching across Perú and Brazil, splitting into numerous distributaries and channels, before opening up into the Atlantic Ocean. How many water systems need to be considered when talking about protecting Amazonian pink dolphins? How many floodplains give life to the delicious camu-camu, serving as starting points for conservation strategies? For a split second, I dreamed of traveling the entirety of the river and figuring out the answers myself. I wouldn't be the first.

The Amazon water system represents an extraordinary challenge for any navigator daring to cross one of the longest rivers in the world; it stretches for 6,400 kilometers (4,000 miles) and is second only to the Nile River (6,650 kilometers, which is about 4,100 miles). Notable explorers include Francisco de Orellana, the first European to navigate the entire Amazon River, in 1542, and Ed Stafford, who trekked its entire length on foot, finishing in 2010 after more than two years. It is reasonable to imagine that local Amazonians crossed the entirety of the river many years before these expeditions, although a lack of written records makes it impossible to know.

The length of the river is sometimes debated, depending on whom you talk to and the metric references used to calculate a river's origin and farthest point. Generally, a river's source is considered the most distant point in the river's longest tributary that flows continuously and without disruption all year round. However, as some scientists argued with newly collected data about a decade ago, if we expand this definition to include tributaries that flow intermittently—meaning they dry up for some periods of the year, whether due to natural or man-made causes—then the Amazon River would become the longest river on Earth, surpassing the Nile.

Interestingly, this is not the only point of debate. For a long time, there was a dispute about the origin site of the Amazon River, with a dozen water sites competing for the title. The dispute continues to this date in some circles. However, in 1971, a National Geographic expedition identified a lake sitting at the base of a snow-capped mountain peak in the Andean Mountain Range in Perú as the headwaters of an important tributary of the Amazon, the Apurímac River, earning it the title of the *true* source of the Amazon. The mountain peak identified is known as the Mismi, located in the region of Arequipa, and it harbors a cliff that dispenses glacial water into the lake that is recognized as feeding the most likely source of the Amazon.

As a teenager, I traveled to the region of Arequipa, trekking into high-elevation mountains with peaks 4,000–5,000 meters (13,000–16,500 feet) above sea level. After a lot of physical effort, feeling as though my lungs were in a compressed box that kept getting tighter, we reached a lake with crystalline blue waters. Behind the lake, tall pointy mountains perfectly reflected in the still waters created a hypnotizing imagery that became etched into my memory.

Although it was not obvious to me then, as an adult, I recognized that somewhere between those mountains lay the origin of the Amazonas— the river I was now traveling for scientific exploration.

Despite all the scientific debates that have encircled these waters, what is not disputed is the fact that the Amazon River is the largest river on our planet by water discharge, contributing 20 per cent of all the freshwater that flows into our oceans. By this point, the waters have carried sediment and other microbial life from most of the continent, mixing it with the ocean to provide unique nutrients that support marine life.

I find it wild to imagine that glacial water and raindrops, first accumulating in the high Andes of Perú, flow and cut through rock and land structures naturally formed over millions of years, cascading down steep slopes, forging rapid streams and gorges. As the water descends, the river grows in volume and force, carving the landscape. As the altitude lowers and it reaches the lowland floodplains, the river's flow slows down, gently growing broader and depositing sediment transformed from the

Andes, enriching the soil of the Amazon Basin, eventually reaching the Atlantic Ocean.

As I often say, what happens in the Amazon *literally* impacts everyone around the world, whether you're in England, Australia, or anywhere else on the globe. If the intense periods of drought in the Amazon persist or worsen, a phenomenon largely driven by deforestation, the amount of freshwater captured by the Amazon River to enter the world's oceans would reduce. This, in turn, could impact ocean currents, salinity levels, and global climate, potentially disrupting marine ecosystems and the lives of the humans who depend on them.

Enjoying the sounds of our Amazonian boat crashing against the water, I lay down on the slender bench, arms crossed, legs stretched out, and closed my eyes for a moment as we continued navigating toward our next destination.

After another 30 minutes, we entered a smaller tributary of the Amazon River, finally arriving at our destination for the day—a community deeply connected to these waterways for centuries, relying on these rivers for food, communication, transportation, and culture.

The Yagua.

I hadn't visited the community since I was ten, and we were here to greet them after a long time and learn more about their indigenous relationship to water and all life forms that depend on the river.

As we climbed the tall riverbank, our boots began sinking into the mud, forcing us to hold ourselves up using the wooden poles scattered around, playfully competing to see who could ascend the hill faster. I wiggled my rubber boots left and right, tensing my toes to create an anchor that helped push my boots upwards, contributing to the force needed to hoist myself up as I pulled forward with all my core strength.

Following the trail to reach the *maloca*, the community's traditional communal house, we encountered a Yagua man leaning over a red bucket filled with water and fish. He was dressed in a traditional long skirt made from palm fiber (*Mauritia* sp.) with a natural beige tone and splashes of reddish-orange dye. His head was adorned with a piece made from similar materials, with longer fiber pieces stretching back, and his arms were adorned with braided armbands with dangling fibers, complemented by strokes of red paint on his cheeks.

The man greeted us with a smile and a gentle nod of his head. He was meticulously cleaning each fish before placing it in a handcrafted thin wooden bowl, intricately carved on the outside.

The Yagua people, an ancient indigenous culture—and one of the largest in the Peruvian and Colombian Amazon, with close to 6,000 members alive today—are highly skilled fishermen who live near the Amazon River and its nearby tributaries. Many still employ spears to catch fish in rivers and lakes with remarkable precision. The Yagua are intimately familiar with the feeding and reproduction habits of every fish they consume, caring for the ecosystems and determining the optimal times for hunting, in alignment with the ecological cycles of local water bodies. They balance artisanal fishing with rotating crop structures to preserve biodiversity, ensuring they take no more than is needed and preventing waste.

We arrived at the center of their community, a cleared space of land surrounded by lush vegetation. There, two male leaders waited for us: Juan and Carlos. The shorter man, Juan, wore a headpiece featuring large colored feathers at the front, and was holding a large blowpipe in his right hand. Carlos's headpiece was thicker in palm fibers but only boasted two smaller feathers atop. Like the man we'd seen earlier, each had two thick lines of red dye on their cheeks.

The red ink was achiote, also known as annatto (*Bixa orellana*), a potent tropical healing fruit from the Amazon. It's also one of my go-to Amazonian products for almost every use, from cooking and make-up to UV protection and more. Conical in shape, this fruit has a hairy capsule that opens to reveal anything from 20 to 50-plus tiny red seeds. These seeds are integral to Amazonian culture. The natural red dye is used by indigenous communities like the Yagua as face paint, and is applied daily or during cultural ceremonies and rituals to indicate hierarchy or to embrace the energy and abilities of other species. The face paint may also serve as a defensive mechanism to convey fear and protect from danger, or to express moods and feelings.

My favorite use—a tip I learned from Amazonian women—is using achiote as a natural lipstick and sunscreen (the tree is, in fact, sometimes referred to as the "lipstick tree"). The fruit is also utilized in cooking and medicinally, to treat asthma, diarrhea, and other conditions. The chemistry embedded within this fruit is very powerful. Interestingly, annatto is already widely commercialized worldwide as an industrial food coloring for butter, cheese, ice creams, and meats—so you, too, have probably consumed it at some point.

My grandmother planted an achiote tree in our backyard a few months after I was born. It stands tall in a corner, providing shade and a ready supply of cooking seeds. In fact, ever since the fruit began to flower, my grandmother has ceased using paprika and loyally uses achiote, making for delicious chicken.

After exchanging a few words and warm hugs, Carlos guided us to the *maloca* for their welcoming ritual before we proceeded with further interactions. There, he asked us to sit as he painted our faces with two strokes of achiote on each cheek while speaking some words in Yagua, thanking the waters for bringing us safely.

Following a beautiful dance, our Yagua brothers invited us to try our blowpipe skills. A medium-thick cut trunk laid against a wooden pole stood at the corner of the cleared ground outside the *maloca*. Carved into the trunk was a semi-deformed animal face at eye level, symbolizing a hunter's skill. Holding the long, heavy blowpipe in one hand, Carlos placed a dart on the other end.

When hunting, the dart is dipped into a very poisonous substance known as *curare*. In Latin or Spanish, *curare* means "to heal," which is somewhat ironic given that this natural cocktail is used for paralyzing prey.

When European conquistadors entered the Amazon, they encountered warriors who blew darts at them. When hit, the soldiers would fall from their horses, their bodies immediately entirely paralyzed. Some tales were spread that this poison was prepared by powerful witches who lived on the outskirts of communities, spending days preparing the concoction in giant cooking pots. Supposedly, the cocktail was ready when the women would fall unconscious from smelling the toxic fumes emitted by the poison. These were false rumors.

Each community has their own recipe for *curare* preparation, with two plants as the main components: *Strychnos toxifera* and *Chondrodendron tomentosum*, sources of the toxic molecules toxiferine and alkaloid D-tubocurarine. Sometimes, ants or snake venom are added to the cocktail, augmenting the toxicity of the preparation. Interestingly, this poison is only active when it enters the bloodstream directly, remaining inactive if ingested orally.

Carlos placed a palm fiber headpiece on my head as he tapped my shoulder in a sign of good luck. I positioned my left leg forward and held the blowpipe firmly with my right arm, pressing it against my mouth. I took a deep breath and aimed upwards, laser-focused on the target—the forehead of the carved animal. In a single, swift motion, I expelled all the air from my lungs, watching the dart travel speedily to hit the target.

The Yagua cheered loudly, delighted that I still remembered what they had taught me years before, inviting me to come visit them again soon.

We spent the rest of the afternoon together, chatting and learning about the Yagua's deep connection to water and the life it sustains. They shared that it is now difficult to catch fish in the nearby waters. The fish are no longer coming regularly. This phenomenon, influenced by climate change, contamination and commercial fishing exploitation, is putting their traditional knowledge and food security at risk, forcing them to find alternative ways of earning income to purchase fish at the nearest city market. A painful irony. When they do find some native fish, as they had done that morning, they honor the fish deeply for nourishing their bodies. Their relationship to the river and its natural resources is crucial to their culture and survival. We purchased some of the Yagua's fine and beautiful handcrafted art, including masks made with dried wood to accompany the collection I already had back at our home in Perú. Then we said goodbye, promising to return soon as we thanked them for their generosity and time.

Navigating the Amazon River and its tributaries for an entire day left me feeling invigorated and profoundly calm: a delightful mix of sensations that only nature can provide.

Witnessing wildlife thrive in its native habitat, peacefully accepting our presence, was humbling. It was equally enlightening to observe the balance between flora and local communities, where people sustainably harvest only what they need for survival—no more, no less.

In the Amazon, water connects everything, from the faintest drop of rain in the high Andes to the deepest point in the Atlantic Ocean. It nurtures the fertile soils that foster native trees, and supports the flourishing fruits that provide nourishment to its inhabitants, from the reputedly ferocious piranhas and mythical Amazonian pink dolphins to the joyous human communities like the Yagua.

Globally, water bridges people, culture, and wildlife, transporting minerals, nutrients, microbes, plants, and animals. It carries endless genetic information and a wealth of stories, practices, myths, and tales. This is especially true for the Amazon River and its hundreds of tributaries, which weave through nine countries in a flowing pattern, merging their destinies with each twist and turn, in a constant dance from the mountains to the sea.

The Amazon River and its myriad waterways aren't only important for local and global ecosystems; they are also central to Amazonian culture, and to the countless life forms and traditions that originate from, depend on, and flourish because of these waters. Many secrets still lie within them, waiting to be discovered.

As I observed the sun set over the Amazon that day, I couldn't help but feel how deeply intertwined our lives are with these ancient waters. It is this interconnectedness that strengthens our commitment to ensure the Amazon continues to thrive—not only for the health of those who live along its banks, but also for the health of our entire planet.

4

EVIL SPIRITS

Amazonian tobacco

SCIENTIFIC NAME: *Nicotiana rustica*

TRADITIONAL NAME: Mapacho

ORIGIN: Native to the Andean and Amazonian regions of South America

TRADITIONAL USES: Amazonian tobacco is better known as *mapacho*. This plant, *Nicotiana rustica*, is different from the well-known tobacco consumed in cigarettes and other forms, a variety known as *Nicotiana tabacum*. *Mapacho* is considered a powerful healing medicine that works spiritually and energetically, cleansing the energy, removing spiritual blockages, and inducing an altered state of consciousness. This tobacco is often used to blow *"mapacho* smoke" on participants of ceremonial rituals and events, to facilitate communication with spirits. Many Amazonian communities use

Amazonian tobacco when hiking in the dense jungle to ward off the evil spirits of the rainforest.

SCIENTIFIC INFORMATION: *Mapacho* contains a higher concentration of nicotine, a naturally occurring alkaloid with psychoactive properties, than conventional tobacco, making it an important and potent ingredient in traditional practices. This herbaceous plant can reach heights of up to 1.8 meters (6 feet) in optimal environmental conditions. It has elongated, trumpet-like yellow or white flowers that are fragrant upon blossoming.

I felt the ground's proximity as the vehicle dipped and rose on the rough terrain, jolting and shuddering its way through one of the bumpiest rides I'd ever experienced. The non-stop vibration of the roaring engine resonated through my bones, from my coccyx to my jaw. We were zigzagging through one of the busiest cities in the Peruvian Amazon, the wind pressing against my cheeks, tugging at my face with an invisible hand. From the corner of my eye, I saw my mom sitting next to me, bracing herself against the cushionless seats and gripping the poles to maintain her balance.

– Maestro, *can we slow down just a* . . . My words were drowned out by the engine's rev as the wheels whirred in a deep rumble. The driver veered to the right abruptly. Metal clanking, horns honking, *mototaxis* darting all around as street vendors shouted from the sidewalk.

A *mototaxi*, the preferred mode of ground transportation in the Amazon, is a small, motorized vehicle with two seats in the back that can navigate the winding and uneven roads of some of the most remote areas of the rainforest. In other parts of the world, such as India, the vehicle is known as a tuk-tuk. Often painted in vibrant colors and personalized with flamboyant decorations, these vehicles reflect the culture and individuality of each driver. They provide a source of income for families—and cost-effective transportation for tourists—and facilitate the exchange of goods and services. Growing up in Perú, *mototaxis* were my version of rollercoasters. And oh boy, did they always deliver. Uphill lifts, sudden drops that turned the stomach, twists between trucks and roads, and even splashes of water—all the elements of a good ride.

– Señorita, *those kids were about to throw colored water balloons at us* – the driver yelled in frustration over the bustling city sounds. *I just showered!*

I turned quickly to see a group of three teenagers laughing mischievously as they sprinted down the street with their faces and clothes streaked in multi-colored ink—purple, pink, blue, green. They ducked just in time as a second group of teenagers retaliated, hurling two giant water balloons over their heads. The balloons crashed against the wall and explosively burst, splattering yellow-colored water everywhere.

– *I'll just keep dodging the balloons* – the driver declared decisively, as a bright blue *mototaxi* sped by, laden with heaps of freshly picked bananas, secured loosely at the back with thin green plastic ropes. *Today is the last day of the Carnaval,* señorita*!*

The Amazon Carnaval, recently declared a national cultural heritage event, stands out as one of Perú's most vibrant cultural and spiritual celebrations. It merges lively music, endless dancing, colorful parades, traditional rituals, and, indispensably, water balloons. Rooted in ancestral beliefs, the Carnaval is held during a time when evil spirits roam free and the portal between dimensions is open. Locals dress as these spirits or as mythological creatures of the Amazon, turning the event into a lively and exuberant experience. Many wear dried seeds around their necks, red paint on their faces, black ink lines and shapes decorating their bodies, parrot feathers in their ears, and tree fibers tied around their waists, along with a variety of costumes created with natural dyes and resources from the rainforest. Some do it for good luck and fun, some to show respect to the Carnaval, and most, without a doubt, to secure protection against evil spirits.

The *mototaxi* turned left, taking us onto a narrow road flanked by traditional food vendors. The aromatic blend of fresh fried fish,

Amazonian herbs, and banana slices was making me salivate. After a few more minutes of jostling around in the back seat, we came upon a series of colorful tents stretching several blocks ahead and to the left. The sound of the tires shifted, crunching against gravel as we neared one of the town's most popular street markets. The *mototaxi* came to a halt, and the driver offered us his number for a return journey. Hair disheveled, eyes watering and heart happy, I smiled politely, pocketing his number but making a mental note to walk the rest of the journey. My mom laughed uncontrollably as I helped her exit the vehicle. Talk about a lineage of strong women.

Stepping into one of the Amazon's most renowned traditional medicine markets, we found ourselves enveloped by vibrant ancestral wisdom. The space was teeming with life. Our mission was to document the most commonly used natural remedies and grasp the impact of these Amazonian resources in local health, economy, and culture.

Serving as a natural pharmacy, this market represented an important point of commerce and social life for locals. All around us were people and movement, from men balancing sacks of rice on their head while greeting friends and vendors, to women carrying bowls of fish and medicinal herbs while exchanging the latest town gossip, and children playing cards and marbles using ancient, dried seeds. I even noticed Peruvian hairless dogs roaming freely,[1] their presence adding an air of mystique to the elixirs of traditional healing and the waning hours of Carnaval. Centuries ago, this ancient breed was revered as sacred, believed to possess mystical powers that could ward off diseases and evil spirits. Everything in this market oozed life, and I took a few minutes to soak it all in before walking down the corridor of colorful tents that stretched along the road.

The traditional market in the heart of the Upper Amazon was full of vendors beckoning us to discover cures for any ailment. Plants to cure infertility, roots for treating laziness, seeds to reduce inflammation, flowers meant to attract love. Tables draped in white or colorful cloths bore bottles of freshly prepared remedies, promising maximum potency and showcasing an array of traditional medicinal plants from the rainforest. I even noticed a seller of *emoliente*, a go-to remedy in Peruvian culture. This traditional potion is prepared by combining herbs and seeds in boiling water along with honey to create a soothing drink that is known for aiding digestion. My eyes darted everywhere, eager to absorb as much as possible, before following my mom over to a vendor to find some seeds for my grandmother.

– *I also have* colmillos *(fangs)* – the woman in front of us mentioned casually in a soft tone as she peeled some native fruits, seeming unbothered about whether we heard her or not.

– Colmillos? I echoed, startled, my eyes quickly scanning her stand in search of bones, finding only leaves and bottles. *Are they real?*

She bent down to retrieve a box from beneath her table. Opening the loosely closed lid, she revealed an array of necklaces crafted from dried tree fibers, each adorned with seeds of white, gray, and brown hues gathered from the rainforest. Caught in the sunlight were fangs, integrated into the traditional jewelry. Their varied sizes and thicknesses hinted at the multiple species from which these fangs, believed to possess mystical powers, had been extracted.

– *This is from a* jabalí *(wild boar)* – she said, selecting a white, slightly cracked medium-sized fang that dangled from a necklace made with dark-inked fibers. *A caiman* – she continued, showing us a thick, white reptile tooth secured to a necklace with metal.

My mom set aside the seeds she had been examining, and we stared at the display of teeth.

– *And this . . . this is the* tigre – the woman announced picking out the largest fang: an off-white, pearl-like, and pointy tooth, securely tied to the necklace, its weight causing the necklace to hang lower. *Or jaguar, as most people know it. It's an amulet for good luck, to ward off disease and for serious protection against evil spirits* – she explained, her eyes widening and her voice deepening as she placed the jaguar's fang in my palm, closing my fingers around it and patting my hand in a maternal gesture. The fang pinched into my thumb, sending a shiver of electricity down my spine.

In some regions of the Amazon, animal teeth are valued for their protective qualities against malevolent forces, including those that cause sickness. The age-old practice of wearing animal teeth or incorporating them into ceremonial rituals aims to transfer the animal's attributes to the wearer—its strength, agility, and courage. These items serve as spiritual and protective amulets in Amazonian mythology, where certain predators are esteemed for their ability to ward off evil. Unfortunately, over time, this cultural connection is becoming exploited for international commerce, with animal parts being prized for the sake of ego and profit.

– *I also have* isulas *(bullet ants)*[2] *for good luck* – she disclosed, revealing a second box from beneath her table. It was filled with glass decorations and keychains, each holding one or multiple bullet ants, frozen in time.

I picked up a keychain to examine them more closely. Some ants were as long as half of my finger, and I shuddered at the sight of their sting—known to hurt as much as being shot with a bullet. There is a community in the Brazilian Amazon called the Sateré-Mawé, where dozens of bullet ants are placed inside a pair of gloves for a rite of passage. To transition into manhood, young men must wear these gloves for at least ten minutes, enduring this trial repeatedly over the course of a year. This tradition honors the profound power of animals, while showcasing a young man's resilience and capacity to withstand pain, indicating his readiness for responsibility and leadership within his community. A tingle ran through my fingers as I gently placed the ant ornament back in the box.

Each ant decoration or keychain was accompanied by a *huayruro*,[3] a seed native to the rainforest, typically red with a black spot, said to protect against the *mal de ojo* (evil eye) and bad luck. The evil eye is a belief widespread across many cultures globally from the Amazon to Europe, Africa, and Asia. The earliest known references were found in ancient Mesopotamia, carved in cuneiform tablets dating back to about 5,000 years ago. The evil eye is a malevolent gaze, rooted in envy and dislike, that is given to someone and causes injury, sickness, and misfortune. One of the most well-known protective amulets against the evil eye is the nazar in Turkish culture, an eye-shaped amulet that is made of glass and composed of circles or teardrops in blue, white, and sometimes black.

The contrasting colors of the *huayruro*, red and black, are believed to represent duality and balance that are key to *protección*, as my grandmother would often remind us. Children and adults alike often wear jewelry made from *huayruro* seeds, including bracelets, necklaces, and earrings, and may even decorate their homes with crystal jars full of the seeds.[4] In South America, the evil eye is particularly feared for newborns and young children who are thought to be the most susceptible.

THE SPIRIT OF THE RAINFOREST

I touched the *huayruro* bracelet on my left wrist, recalling the first time I began wearing a *huayruro*. I was only six. My cousin Rodrigo had suddenly fallen ill. Two years older than me, Rodrigo was my faithful partner in crime; we would stage mud fights together, proudly sing Shakira songs using kitchen ladles as microphones, and learn how to play the latest video games. He was unable to retain any food, and after seeing one expert pediatrician after another, my aunt was inconsolable. One doctor suggested surgery, another hospitalization, and one that we all join together in prayer, hoping for the best. No medical test could pinpoint a bacterial infection, chronic disease, or any identifiable condition that could be addressed directly. At the time, I couldn't fully grasp what this meant; I could just sense the stress and sorrow of those around me, and felt confused as to why Rodrigo could no longer play with me.

That was the first time I heard the word *tunchi*: the spirit of a tormented man that now wanders the rainforest like a lost soul searching for redemption—or rejoicing in terror, as most describe it. This spirit is said to quietly approach those who are alone, drawing near enough to emit a high-pitched whistle, a harbinger of death for those who hear it. Resembling the singing of a bird, it is only heard at night. I've been told by Amazonian locals that sometimes they hear its call from kilometers away, across the expanse of the jungle; they may mistakenly think it has departed, only for the whistle to resurface suddenly by a riverbank or above a house. Legend has it that persistent singing by the *tunchi* means bad events are approaching, such as sickness or death. One must never acknowledge its call, for responding invites the *tunchi* directly to you. My uncle, who had lived in the rainforest for many years before relocating to Lima, speculated that perhaps Rodrigo had been cursed by the *tunchi*.

EVIL SPIRITS

It wasn't until we met Don Pablito, a renowned traditional healer, that things turned around for Rodrigo—or so our family believes to this day. Welcoming us into his adobe house on the outskirts of Lima, Don Pablito, who spoke only a handful of words in Spanish, began to pray in his native tongue. He methodically rubbed a freshly collected hen's egg around Rodrigo's head, upper body, and stomach. A South American tradition known as "passing an egg," this ritual is meant to remove illness or fear from a person. I vividly recall the scent of adobe mingling with the aromatic Palo Santo smoke as I quietly watched what unfolded.

Don Pablito then cracked the egg into a tall glass of clean water away from Rodrigo, "reading" the egg to diagnose the problem. After several sessions, he declared that Rodrigo had been afflicted by the evil eye, but the eggs, one now cooked, were absorbing the negative energy. He asked the family to remain patient and hopeful, as Rodrigo would soon recover his health. The exact cause of his illness and the means of his recovery remain mysteries to us. However, what is certain is that Rodrigo is now strong and healthy, and these experiences gave me a closer affinity with my roots. Ever since then, my grandmother has insisted that we wear *huayruros*. While we might not always remember, I make it a point to carry one with me, be it on my bracelet, in my backpack, or just placed inside my closet at home. I've grown up using the seeds, worn as necklaces or bracelets, and they are now so ingrained in my cosmovision that I instinctively wear red and black to important meetings or interviews for good luck.

We thanked the woman for her time, purchased a kilo of my grandmother's inflammation seeds, and continued on our way. These ancient practices serve not only to protect against malevolent spirits that might whisk one's soul to the unseen corners of the Amazon, never to return, but also to shield against negative energies that could harm one's business or bring sorrow and pain to one's family.

Actions rooted in tradition can speak louder than words, emphasizing that our efforts to explore and protect the natural world must not overcast or endanger these ancestral practices. This is how we honor our planet's biocultural heritage.[5] In fact, traditions, just like the Amazon Carnaval, can be a rich source of learning, empathy, and joy—celebrating the cultural expression of nature where spiritualism is inherently embedded into every aspect.

Before the sun set for the day, my mom and I joined the rest of our group and we hopped onto a *peke peke*, which fearlessly bobbed through the waters of the Marañón River. The motor worked tirelessly, singing "pek-pek-pek-pek" as we traversed the characteristically brown tributary of the Amazon River. Maneuvering around several fallen tree trunks, we made our way to the riverbank and pulled up to the entrance of the Northern Amazon community we were visiting.

We were welcomed by a man called Heriberto, a leader deeply rooted in Kukama-Kukamiria culture, along with his beautiful wife Rosa and their entire family. We unloaded our bags, tents, and mats, which had been carefully placed in the *peke peke* with Tetris-like mastery to avoid anything getting too wet. The Kukama-Kukamiria is an indigenous culture that is deeply intertwined with the biodiversity of the Amazon, carrying profound knowledge of the rainforest, and maintaining a strong bond with the spiritual world. The children greeted us with open arms and joyful chatter, visibly excited about us staying with them for the next few nights. We were here to learn about how the Kukama-Kukamiria identify wildlife at night in a practice that is key to their production of medicinal and spiritual remedies, and dates back to Heriberto's ancestors, who founded this community. We aimed to use the collected data to advocate for the conservation of their culture and biodiversity.

As we trekked through the jungle en route to Heriberto and Rosa's home, which was located 30 minutes away from the riverbank, the setting sun cast a golden glow that filtered through the trees, adding to the breathtaking beauty of these lands.

Chris, who was walking next to me, had taken on the color of a ripe tomato and was sweating profusely thanks to the rainforest's generous humidity. In his right hand, he carried a bag; above his head, three mats were rolled together and held aloft. He was covered in fabric from head to toe, with long sleeves, a neck cover, thick socks, and tall rubber boots that marched forward through the slippery mud that stretched between the underbrush and trees. I could only see his piercing blue eyes and nose. Just a few days before traveling, we'd met other British people in Perú, who'd teased him and warned him of the *izango*—a sneaky bug that bites you on exposed skin and deposits eggs, which had happened to some of them during their first trip to the rainforest. Due to COVID restrictions, visa paperwork, and the demands of his own work, this was one of the first times Chris had been able to come and explore the Amazon with me, and he was committed to avoiding the *izango*. A lad from West Brom in England plodding through the Amazon with his Peruvian wife: I would have paid to watch that on TV.

We were in safe hands with an expert team of explorers. Cesar, an invaluable field-research partner and a descendant of the Kukama-Kukamiria, managed the bottles of water and hammocks. Luis, a local biologist and skillful photographer, carried the rest of the equipment. We were ready to adventure into the depths of the surrounding territories, familiarizing ourselves with the space, flora, and fauna as we scouted the area for a scientific study.

We all gathered in Rosa's cozy kitchen, where a rustic fire crackled and popped, casting a warm glow over the crowded room. A generous pot of chicken soup simmered on the stove, filling the air with a delicious smell. While eating dinner, we caught up on the past year's adventures and misfortunes, like a long-lost family reuniting again. I had visited Heriberto and Rosa many times before, collaborating with them to unite science with conservation efforts on the ground, and we'd developed a strong friendship over time. Outside, crickets chirped loudly through the traditional window, a square hole in the wooden wall. With the last remnants of power-generated energy, we decided to set camp before it got too late.

As we started gathering our equipment for the night trek, someone entered the room. The team erupted in cheers, greeting the newcomer with excitement. Before I even turned around, I smiled, already knowing who it was: Manuel. I felt an overwhelming mix of happiness and relief at knowing he had agreed to join us. Manuel navigates the Amazon at night like no one I have ever met before. Though we had only met him the year before, he already felt like part of the family.

I remembered the first time we'd met Manuel. He had heard some scientists were staying at Heriberto's hut, and thought we would be interested in his latest catch. It was 5am and my eyes were still shut when I heard the growing clamor of curiosity just outside; something had attracted the kids' attention. I stumbled my way out from under the mosquito net, rubbing my eyes and searching for my shoes.

– Doctora, *I thought you might want to see this.* I heard Manuel's voice as I stepped outdoors to join the excitement.

He was holding the head of a freshly caught, medium-sized blackish snake with thick, pointed teeth. Its luminous scales gleamed under the rising sun, and a forked tongue protruded, which would have allowed it to collect odor molecules simultaneously from the air and ground.[6]

– *Morning* – I said, my voice coming out raspy in such close proximity to the snake's head. I saw Chris and my mom looking stunned right next to me.

We'd soon learned that Manuel was known in the town as the "night hunter." When no other work was available, he felt forced to catch exotic animals and sell them in local markets to earn some cash and feed his family. He explained that snake heads were always in great demand. Local culture believes that snake heads bring luck to new businesses when placed inside a jar in the corner of an office or shop. Manuel would likely get 300 soles (about $100 USD) for this catch. Before the COVID-19 pandemic, he had been part of a reforestation program in the buffer zone[7] of the Northern Amazon. However, oversight and a lack of follow-up had left many Amazonians unemployed, without severance pay or any support to find alternative work. This had led him back to what he knew best: finding animals at night.

Amazonians have a unique relationship with the animal world. Ancestral cultures believe that animals have a human-like spirit, carrying the deep wisdom of the rainforest. In this worldview, animals not only experience joy and pain but also embark on a spiritual journey. Thus, interacting with animals transcends the physical realm, entering a space where all spirits are equal, living in reciprocity. We discovered that Manuel is a descendant of the Kukama-Kukamiria, and that he experiences pain when hunting animals. He believes that hunting for economic benefit harms the spirits of animals, and that this pain reverberates beyond the visible world, perpetuating a cycle of sadness in the spiritual realm, and sometimes hunting him in his sleep. He seeks permission from the Pachamama and the residing spirits, explaining that he only resorts to hunting when his family has no other means. His understanding for the need to protect local ecosystems in the Amazon is deep and vast, as is his empathy for nature and the life forms that call this place home. But just as honest and real are his needs to provide for his family in the crude reality.

Back in Rosa's kitchen, our briefing began.

– It will be pitch black and we will be in virgin jungle – Manuel announced. *You focus on your science and cameras. I will look for wildlife. I have encountered some* gatos *(cats) in these lands before.*

My mother squeezed my hand, her gaze fixed directly on me. Suddenly, the mosquito repellent on my skin began to sting.

– Tonight, you trust me fully. You trust me without hesitation. If I stop and blink my light three times, you must stop and immediately stand behind me. Quiet as a rock, ready to move on my command, as it may be a predator—and I don't mean animals only. Got it?

Everyone nodded in agreement. We all understood the risks involved.

Rosa kissed Heriberto on the cheek. Luis agitatedly rummaged through his bag to find his spare batteries and sighed in relief. Manuel and Cesar grabbed the machetes and headed outside. Chris was already waiting for me by the door. I approached my mother for one final hug, and I felt her hand slip something into my pocket: garlic, salt, and Amazonian tobacco, just in case, as she often did before my journeys into the jungle. She swiftly made the sign of the cross on my forehead[8] and asked me to return quickly.

EVIL SPIRITS

Roxana, Heriberto and Rosa's older daughter, rushed over from a corner and reminded me to wear *la muñeca* around my neck: a small doll with a playful pink dress and long black hair, all dyed using medicinal seeds and fruits. Roxana and her sisters had lovingly crafted this doll for me earlier in the day, noting that she looked like me, but also like them. They explained that *la muñeca* would protect me from the evil spirits in the jungle—because like them, I, too, was a guardian of the Amazon.

For nearly an hour, we trekked through the winding paths of towering, shadowy trees and glistening bushes, illuminated by the full moon. Manuel and Heriberto fearlessly forged ahead, clearing a new path with their machetes, unbothered by the lianas tripping their boots or the scorpions skittering nearby. Croaking, whistling, rustling, buzzing. The night was alive. As the stars gradually disappeared behind thick clouds, we found ourselves admiring the majestic trees that emerged on this new path with fascination. This was my first time conducting a night exploration in this area. Heriberto had shared with us that among the plethora of natural resources in the rainforest, there are some animals that produce unique substances to cure illness and honor spirituality—and that you can only find them at night. I was eager to see the wildlife that emerged within the depths of these lands.

It was during this enchanting night trek that Heriberto shared with us the story of the time he encountered the fearful *chullachaki*.

A mythical creature whose name derives from the Quechua language, the *chullachaki* ("choo-ya-cha-kee") is often described as a small, ugly spirit with one leg shorter than the other, and one foot larger than the other. Legend says that it lives deep in the Amazon Rainforest, and that it takes on the shape of a loved one to persuade its victims to follow it deeper into the trees until they cannot find their way home. It is also believed that the *chullachaki* can take the form of any animal in the Amazon. It is regarded as a guardian of the land that punishes those who act unwisely or break a taboo.

– *Everything went cold* – Heriberto recalled. *I felt it in my bones, as if it was staring at me. He started to call my name. I could hear it so clearly, it sounded like my kids calling for help. I ran, as far as I could and as fast as my legs could go. I just ran. But you know there is something wrong when the jungle goes silent. It feels as if someone turns off the radio, completely, abruptly. And for as long as I have lived in the rainforest, that has never happened, especially at night when everything comes alive.*

Heriberto explained that the *chullachaki* has taken people before with no trace. Although he had never seen it directly, he knew of people who had. Descriptions varied, with some depicting it as an elderly man wearing tattered clothing and a long hat, while others swore its left leg resembled an underdeveloped, chubby baby's toe. Despite the different accounts of its appearance, one fact remained consistent: the *chullachaki* was merciless, and venturing near it was not to be risked. Instinctively, I squeezed the garlic, salt, and Amazonian tobacco in my pocket, and touched the protective doll around my neck.

We kept advancing, breathing in the fresh, pristine air of untouched jungle: my favorite smell in the world. We trekked in silence for another 30 minutes, stepping over lianas and rotten tree roots, absorbing the beauty and the smells of the rainforest at night.

Without warning, Manuel stopped and clicked his flashlight three times, his body motionless, with one arm extended into the air. My body froze in response.

His eyes meticulously inspected the jungle around us. Something on the right side of the path had seized his attention. It felt like we were all frozen in time, paralyzed, avoiding any further noise. I thought of my mom waiting for us in the hut. I thought of my grandmother back in Lima, oblivious to the adventures and troubles we had embarked upon. I thought of my husband, who stood just a few meters away from me.

The cerebral cortex of my brain reminded me to listen closely. I could still hear the sounds of the living jungle around us; it wasn't on pause. Deep instincts rooted in my gut and ancestry told me this couldn't be the end. It simply didn't make sense. But the rush of adrenaline and cortisol had tightened a knot in my throat, and I felt the flight-or-fight instinct at its core.

Suddenly, the pocketknife in the right pocket of my cargo pants felt hot. My memory ignited like a flame, retracing every step we had taken to reach this point. The rubber boots might hinder us when running uphill through the mud, but we could shed our equipment to lighten our load. If we banded together and displayed our strength, we might be able to scare away whatever was lurking around, buying us just enough time to return to camp.

I looked at Manuel's face for a sign of reassurance—a confirmation that my instincts were correct and that this wasn't the end. After what felt like an eternity, the muscles around his eyes relaxed, and his expression lightened. He tilted his head, signaling for us to keep moving. It was safe. But I knew we had to remain vigilant. The rainforest is home to feline predators and unexplainable encounters, like the bone-chilling *chullachaki* or *tunchi*—especially during the last few hours of Carnaval, when many spirits are thought to roam free.

*

The experiences of our night exploration reminded me that while we were here for scientific research, there are some things on our planet that cannot be seen or explained. There is a realm of the spiritual that is intertwined with ancestral ways of discovering and connecting with nature. It is this wisdom that keeps calling me back, urging me to return home, explore further, and fight for its preservation.

When the sun finally emerged, before most had exited their huts, I found myself yearning to see the *curandero*, the healer in the nearby town: a wise man or woman, much like my grandmother, who was sought after by all to heal any disease or illness, whether it was in the body, mind or spirit. So Rosa took me to meet Jacinto, the local *curandero*. We found him resting in his blue-striped hammock, breathing in the freshness and quietness of the morning. As he rose to meet me, moving slowly as if calculating every step, his eyes pierced through mine, as if looking further than most can. He nodded, seeming to already know why I was there, and simply pointed at a nearby chair. Without uttering a word, he guided me with subtle yet distinct motions, indicating that I should remove my shoes and socks.

Not entirely certain of what lay ahead, and fully aware of the line of fire ants marching nearby, I took a deep breath and settled into the chair, observing Jacinto unfurl a roll of Amazonian tobacco. With a swift flicker of his lighter, he ignited the gentle heat needed to start awakening the hidden powers within the plant. A peculiar fragrance started to swirl, along with a gush of smoke that only grew in intensity.

Amazonian tobacco, the ultimate remedy against all evil, has been used in the Amazon for centuries to assist shamans and healers in staying safe while connecting with the natural and spiritual worlds.

When European explorers first arrived in Mexico, Central and South America, they witnessed the use and significance of Amazonian tobacco or *mapacho*. They admired the strong effect it had on people and its role in social interactions and rituals, and they took the plant back home for recreational and medicinal use. In fact, Amazonian tobacco was among the first New World plants introduced to Europe.

Amazonian tobacco is different from your common carton of cigarettes. Different species, different taste, different nicotine levels. In fact, *mapacho* was too strong a taste for Europeans. It wasn't until a cousin plant was brought from South America to Europe soon afterwards that the commercialization of tobacco skyrocketed, leading to the global adoption of cigarettes.

The preparations of *mapacho* vary between communities and cultures. Some Amazonian cultures dry the plant leaves under the sun and roll them into cigar-shaped pipes they can light up and smoke to call in protection against evil while promoting healing and enabling contact with the spiritual realm. Others steep the leaves in boiling waters, crushing them with rocks to create a strong decoction that serves for deep cleansing of the soul and energy, and is used to treat mental health conditions. I have also encountered some communities that apply the leaves directly on the skin to prevent infections in open wounds. One of my favorite personal uses for Amazonian tobacco is as a powerful insect repellant.

Standing beside me, Jacinto placed his left hand gently on top of my head. Murmuring words in a native language unfamiliar to me and inhaling from his roll of Amazonian tobacco, he proceeded to blow the smoke directly toward my head, focusing on harmonizing the energy within. This rhythmic and deliberate movement was repeated several times. I could feel the warmth of both his hand and the smoke, causing the muscles in my neck and around my eyes to instantly relax.

– Protección. It was the only Spanish word Jacinto uttered, raising his roll of Amazonian tobacco. He then continued the ritual, blowing tobacco smoke over my back, hands, and feet—areas where energy tends to concentrate, as Rosa later shared. The smoke serves to blend the auras of the patient and the healer, allowing the master to impart stability. This traditional technique is essential in treating several physical illnesses in the Amazon.

Known as *soplada*, this ancient ritual uses *mapacho* to cleanse, balance, and heal a person's body and energy, safeguarding the spirit from malevolent forces. When feeling unsettled or anxious, this technique offers protection and reassurance. According to Amazonian indigenous wisdom, the potent properties of *Nicotiana rustica*, the plant from which Amazonian tobacco is derived, possess the ability to purify and shield by facilitating communication with the spiritual realm.

Feeling like I had just come out of a profound meditation, it took me a few minutes to open my eyes and gather the strength to put my socks and shoes back on. The aroma of Palo Santo now wafted from a corner, part of Jacinto's ritual to rebalance himself after healing a patient. Navigating the jungle at night and hearing first-hand tales of encounters with evil spirits had left me in tension. The emotions I'd felt while experiencing the Carnaval—the embodiment of the ever-present connection between nature and spirit in the rainforest—were also still in my body, a strong reminder that we cannot remove culture from nature, or nature from culture. They belong together, and have done so for centuries past, as is gracefully shown by the people and stories around us. Remembering this interconnection can make us feel more human.

Jacinto's murmur of *"protección"* had reminded me of the many times my grandmother had rubbed medicinal herbs on my hair, shoulders, and arms to cure me from the evil eye, or to calm any stress that might be sickening my body. He had brought me the same sense of peace I felt at home. I thanked him for sharing his wisdom and time, for helping me to brush away lingering thoughts of the *chullachaki*, and for injecting me with energy to continue exploring the spiritualism of the rainforest. Rosa gestured that it was time to return home for breakfast, and, slowly, we made our way back together, enjoying the breezy morning in the rainforest.

Looking at my past and the adventures that lay ahead, I saw with clarity the experiences, beliefs, and curiosities that had brought me here. I also believed, more strongly than ever, that science and conservation efforts must recognize and respect the spiritualism of our lands and communities in order to wholeheartedly elevate and protect our natural world. This union of ancestral wisdom and modern techniques would guide me to my next expedition with Heriberto—in search of bees.

5

STINGLESS BEES

Stingless bee

SCIENTIFIC NAME: Tribe Meliponini (includes dozens of genera)

TRADITIONAL NAME: Varies per region and species (i.e., Maya bees in Central America, sugarbag bees in Australia, *angelitas* in the Peruvian Amazon, *jataí* in the Brazilian Amazon)

ORIGIN: Native to tropical and subtropical forests of Asia, Africa, and Central and South America, including the Amazon Rainforest

TRADITIONAL USES: Stingless bee honey is an essential component of Amazonian traditional medicine, and is used to treat various diseases including infections, inflammation, infertility, and upper respiratory conditions. Amazonians know which bees make the best honey for specific conditions, such as the bee that builds its hive in the soil, whose honey is known to be ideal to treat eye cataracts, or the one whose honey alleviates the symptoms of COVID-19.

Stingless bee honey can be consumed directly or with other foods such as bread, applied topically, or combined with plant extracts to enhance its pharmaceutical potency. The Amazonian wisdom associated with bees has been passed orally from one generation to the next, and now continues to be preserved by the communities that practice stingless beekeeping.

SCIENTIFIC INFORMATION: Research shows that stingless bee honey worldwide contains higher amounts of bioactive compounds and has lower sugar content than commercial honey from the *Apis* sp. (honeybee), opening the door to discovering new lifesaving medicines in honey. The reported therapeutic benefits of using stingless bee honey suggest the presence of medicinal compounds. However, stingless bee biodiversity and their hidden chemistry in the Peruvian Amazon remain largely unknown.

— *We don't know what we may encounter, and we certainly don't know how long we will be.* I paused, taking a few seconds to appreciate the immensity of my words in front of my team. It was finally here. We were about to change the course of science by following in the footsteps of those who had come before us. *We are on the verge of stepping into history, tradition, and culture, all in the name of exploration and science.*

I saw Heriberto's chest expand with pride. He, too, was contributing to science. This expedition was about to become the first documented scientific and visual exploration of the ancient tradition of searching for Amazonian stingless bees in Perú—a practice dating back to our ancestors, those who lived in the Amazon before the Spaniards even entered South America. It exclusively took place at night. During the day, many other insects, like wasps, become active and get busy working, and the sounds intertwine at more complex levels and higher decibels than one experiences in the evening. The practice of searching in the pitch darkness was about immersing oneself in the night, embracing the absence of light and guidance, and acting like bees, simply following their buzzing.

Every time I delve into new scientific topics or explore different areas of conservation, I rediscover how the ancestral practices of our Amazonian ancestors originated: by surrendering to the natural processes, becoming an extension of them. Not by forcing, or imposing, but by letting go in order to find and receive.

Everyone was set to go. I stood up and felt the room quiet down. As I scanned the room, I couldn't help but notice that I was the youngest in the group—not only that, but I was also leading it. As I stood there, about to review our next steps before diving into the darkness of the Amazonia to explore new corners of our forest, I felt fierce, and I felt at home.

I wondered if the Amazons had felt the same. Explorer Francisco de Orellana (see page 64) described legendary warrior women living along the Amazon River who fought at the front lines "so courageously that the Indian men did not dare turn their backs." De Orellana referred to them as the Amazons, captivatingly beautiful women who shared a strong bond of sisterhood, and who were masterfully skilled in archery and as physically agile and strong as men. These fierce and formidable women first caught the attention of the Italian explorer Christopher Columbus during his travels in 1493. In 1524, they resurfaced in a letter penned by the grand conquistador Hernán Cortés, addressed to the King of Spain. He profusely shared that he had just learned about "an island inhabited only by women, without a single man . . . and that many of the chiefs have been there and have seen it." Similar accounts were documented in other parts of the Amazon between 1536 and 1538, in Colombia and Venezuela.

Many tales have circulated about the Amazons, from the tradition of cutting off one of their breasts to improve their archery skills, to the notion of killing male offspring to maintain a pure matriarchy; there are even stories of them kidnapping males from nearby islands for procreation purposes. Whether these tales were products of the explorers' fertile imaginations, or in fact accurate descriptions of ancestral women and their significant role in society, we don't know. Some even theorize that such stories might have been inspired by the Greek mythology of the Amazons, immortalized in *The Iliad*.

In Greek culture, the Amazons were renowned as a group of female warriors and hunters who displayed the same courage and fighting prowess as men, leading extensive expeditions to remote areas of our planet. Their origin varies depending on the myth; they're sometimes portrayed as the daughters of the war god Ares, and sometimes as the descendants of the nymph Otrera. In Homer's *Iliad*, "the women of the Amazon, who fight men in battle" are introduced more than two millennia before Europeans first explored South America. Subsequent ancient texts consolidated this memory and provided more details, such as claims from Herodotus that the Amazons were "slayers of men" who were seen "frequently hunting on horseback . . . in war taking the field, and wearing the very same dress as the men."[1] The myths of the Amazons then expanded into other renowned stories, from the legend of Hercules, in which they play the role of formidable adversaries that Hercules must fight on his mission to steal an Amazon queen's magical belt, to the more contemporary DC Comics featuring Wonder Woman.

The mysticism and perceived reality of Greek mythology may have influenced accounts about warrior women in the Amazon. In fact, it was de Orellana who named the "Amazon River" when venturing down the rainforest, because his encounters with fierce women reminded him of the Amazons of Greek legends. Thus, it is possible that the explorers' classical education shaped their interpretation of the native cultures they encountered, in which they may have seen native women actively participating in war and society. Regardless of what the truth may be, I see the strength of such warriors in the women in the Amazon who are now joining forces to fight for the preservation of our territories.

For centuries, the Amazons were considered a tale of fantasy, an invention of the human mind. However, in the last few years, modern science has unveiled a different story. Bioarchaeology experts have unearthed irrefutable evidence that warrior women were present in the ancient world, including during the Iron Age in Europe. In recent years, archaeologists have made astonishing finds in various locations around the world, including the Isles of Scilly in the UK—skeletons buried with horses, spears, knives, and arrows and bows—that have shattered our preconceptions about historical gender roles. Initially assumed to be male, the marvelous magic of DNA testing has revealed that some of these skilled individuals, interred with the tools of warriors, were, in fact, women. While there is no direct evidence from the Amazon Rainforest, it's important to note that many parts of the region, particularly the Peruvian Amazon, remain largely unexplored by archaeologists compared to other regions, meaning much of its ancient history is shrouded in mystery beneath the rainforest canopy.

The energy in the team was electric as we moved deeper into the impenetrable darkness of the jungle at night. If I were to encapsulate my fondest memories from field expeditions, they would all boil down to this very feeling. It is a sensation that reminds me of childhood, filled with play, discovery, and exploration. Yet, at the same time, we are adults confronting deep fears and contributing to the advancement of science for the benefit of our planet. Over the years, I have learned that this feeling never wavers. Every expedition feels like the first one, whether I am in the Andes, the Amazon, remote lands like Yellowstone and Alaska, or less-traveled areas in Egypt, Vietnam, and China. This sentiment reminded me of the advice I received from some of the great dance artists I learned from during my teenage years.

– *It should always feel like this in your stomach* – I recalled Luna saying, just before we jumped onto stage. I remembered being frightened, nervous, excited, ecstatic, all at once. There weren't enough adjectives in any vocabulary I knew to describe how I felt. At the time I was only 16, and it was my very first dance performance. We were about to perform to a 1,000-person audience, and we'd been working on the performance for hours after school, well into the night, with dark circles under our eyes as a testament to our dedication.

– *The day it doesn't feel like this anymore, that's the day you stop* – she had added, and I suspected she meant it more for herself than for us. Despite having over 20 years of professional experience in the world of folkloric and Latin dance, she, like the rest of us, was on the edge of her nerves.

– *Everyone, remember, we are searching for something we cannot see.*

Heriberto's voice brought me back to the rainforest. After one hour of trekking, the distant streetlights now resembled shining, flickering stars, signaling that we were on the cusp of venturing into uncharted jungle territory. The vegetation ahead of us was so thick it felt like our headlamps couldn't penetrate it. But the machetes were fulfilling their purpose, becoming our guides and opening new paths. Soon, we could begin to breathe in the crisp scent of pristine jungle.

Stingless bees are among the most ancient bee species on our planet, with fossils dating back more than 65 million years, long before any evidence of stinging honeybees.[2] Often underestimated, they are found throughout the tropics of our planet. There are approximately 600 distinct species of stingless bees globally, and nearly half of them have been detected in the Amazon Rainforest, spanning countries such as Brazil, Ecuador, Colombia, and Perú. They belong to the tribe Meliponini, and are often referred to as meliponines. Some stingless bees lack a stinger entirely, while others may have a reduced or a non-functional stinger.

One of the most common questions I'm asked, both by kids and adults, whether in private or on live TV, is: *"Do they truly not sting?"* To this, I respond that while they cannot sting, they can in fact bite, and every time I see people shudder with amazement. As well as biting, these bees have evolved alternative defensive mechanisms to protect their colonies such as secreting specific chemicals or using antimicrobial tree resins to construct their hives and deter enemies. They exhibit reduced aggression compared to honeybees, and are more likely to avoid confrontation and flee when threatened rather than engage in combat.

Each type of meliponine bee boasts a unique shape, size, and color. I have encountered some that possess entirely black bodies and are as diminutive as a grain of rice—so small that I once mistook them for ants. Others exhibit iridescent, almost magical features, with metallic green bodies and wings that shimmer with shades of purple, resembling creatures plucked from a Disney movie. Then there are those that surpass the typical size of a bumblebee; when disturbed, they can deliver a powerful bite that might even draw blood and remove skin.

Most of the world remains unaware that stingless bees exist. In fact, most people I have talked to about them have never encountered them in person, nor seen them in books or on television. This knowledge gap transcends geographic boundaries and extends to scholars, agricultural leaders, local authorities, and beyond.

Notably, early Spanish conquistadores documented stingless bees in their chronicles. These bees were regarded as sacred species and good models of social organization, and were kept in the homes of these communities. Nowadays, some Amazonian cultures keep them for pollination purposes, for their medicinal honey, or even solely as loyal pets. The conquistadores were impressed by the quality of the honey and the good nature of these insects. In various regions of the Peruvian Amazon, some stingless bees are referred to as *angelitas* (little angels), a word implying purity and suggesting a deep affection for nature, which might have been borrowed from Catholicism. In the early 17th century, the European honeybee (*Apis* sp.) was introduced to the New World, including the Amazon, for use in Catholic rituals. By the 18th century, stinging beekeeping had gained such popularity that it eventually outcompeted the less aggressive meliponines. I believe this historical shift is one of the earliest and most significant reasons for the neglect of native stingless bees, and the subsequent dangerous decline they face today.

For centuries in the Amazon jungle, dating as far back as written records reveal, local Amazonians have had a profound understanding of stingless bees. The distinct humming sounds of the bees piqued the curiosity of locals early on, emanating from within hollow tree trunks or buried holes in the soil where these bees construct their hives. Amazonians have cultivated an incredible heightened sense of hearing, with some even capable of interpreting the calls of birds in the jungle to discern the presence of predators or the availability of edible fruits. Certain Amazonian cultures have honed their skills to mimic animals with remarkable accuracy, aiding them in hunting some of the most elusive species in the natural world.

As we continued to forge our way through the thickening jungle, we had no choice but to form a single-file line. The vegetation was growing even denser, and we couldn't risk taking the time to create a wider path. If rain were to start pouring, the journey back to base would become significantly longer, exposing us to greater danger and pushing the boundaries of risk in this expedition.

Heriberto led the way, listening intently for sounds that our ancestors would have followed when searching for wild honey. We all held our equipment tight, our arms, legs and core muscles tensing as we traversed the slippery mud, adapting to the constant changes in elevation as we went uphill and downhill. The evening felt exceptionally humid, perhaps one of the muggiest nights I had ever experienced in the jungle. I couldn't be sure if this was due to the actual temperature and environmental conditions or the heightened anticipation of embarking on this once-in-a-lifetime adventure with our brave Kukama-Kukamiria leaders.

– *Shhh, I think I hear them!* yelped Heriberto with excitement. We had been traversing the jungle for quite some time, our pupils adapting to the darkness, our nostrils delighted with a palette of newfound smells, our legs adjusting to the pace and difficulty of the path. *I think they are the* illotas; *they are very close.*

Did Heriberto just distinguish the species *Melipona illota* based on the sound? We accelerated our pace. We had reached flat ground, with trees as thick as they come, majestically posing all around us. We were following Heriberto promptly, jumping over rotten wood and lianas sneaking out of the ground, tripping up anyone who didn't pay attention. We were quite literally running toward our goal, although most of us didn't know what that was yet.

Zzzzz. Zzzzz. I heard some sounds, but they didn't differ too much from the insects that had been following us around. The anticipation was so high that I started to feel frustrated. Maybe my ancestral instinct for sounds didn't get passed down.

I slowed my pace, breathed in deeply, and started to look around the immense black forest around us. Suddenly, I heard it—a deep, constant buzzing sound, like if low bass mode at a concert had been activated.

It was faint, but it didn't waver. As we got closer, the sound only grew stronger.

– *I think I hear it* – I whispered, my heart beating profusely. The others in the team—Cesar, Luis, Manuel, and Chris—could now hear it too. It was distinct, like a radio had been turned on and you couldn't help but focus on one particular channel. Stingless bees buzzing in the middle of a dark evening in the depths of the Amazon Rainforest, elegantly, confidently, proudly.

Heriberto stopped beside one of the tallest trees we had encountered. With a big smile on his face, he pointed at the buttress flare, where the wide triangular base of the tree transitioned into the vertically growing trunk, which loomed about 2 meters (6½ feet) above our heads. He asked us to pause and maintain silence as he placed his ear near the thick, deep ancestral roots, home to countless spiders, insects, and critters. He then closed his eyes intently with his eyebrows raised, as if he was in deep meditation.

– *Yes,* doctora – he finally said, nodding his head. *I think they are* Melipona illota.

After a minute, he asked us to approach and see for ourselves. He explained that their buzzing pattern, pitch, and frequency sounded like those of the *M. illota* species he had found in the wild many times before. He also raised this *Melipona* species, along with a few others, in his home.

– *They produce delicious, medicinal honey* – he added, nodding his head.

We needed to get a closer view to verify the species, whether it was the *M. illota* species or any of the ten other species of interest for Amazonian traditional medicine. We attempted to use the zoom-in features of our cameras and phones, but the accuracy and lighting were poor, hindering our ability to appreciate the bees' physiology.

Unexpectedly, Manuel climbed the neighboring lianas skillfully, effortlessly climbing 2 meters above the forest floor, hoping to catch a better view of the bees to help us identify the species. He swung his way closer to the tree and jumped. Everyone gasped in astonishment; our eyes could hardly make sense of what had just happened in under 60 seconds. There stood Manuel, holding onto the tree with one hand, his machete in the other, his headlamp casting a glow as he looked around. He leaned forward, having found a tiny, precise hole in the trunk, and observed the native bees flying in and out of their hive. Manuel had assisted Heriberto in building sustainable wooden boxes for stingless beekeeping many times before, making him adept at identifying and working with various *Melipona* species, including the beautiful and highly regarded *M. illota*.

He gently tapped the trunk with his machete to confirm the presence of a complete hive – he had previously found some bees momentarily surviving in smaller groups after being aggressively displaced from their homes.

Although stingless bees are found throughout the Amazon, from lowland jungle to high-altitude rainforest, displaying remarkable resilience to different temperatures and altitudes, they are facing increasing threats. Fires, deforestation, species competition, pests, climate change, and unsustainable harvesting are endangering their survival. These threats are destroying their tree hosts, along with the hives the bees carefully construct over several months. They are also eliminating the vast floral resources that stingless bees need to stay healthy and produce medicinal honey, thus putting at risk traditional medicine and cultural practices.

The courageous *M. illota* bees buzzed louder and more vigorously, ensuring we knew they were fully present and alive. Manuel descended the tree just as effortlessly as he had ascended. I made a note to ask him to teach me how to climb trees in the Amazon later; such a skill may prove useful one day. Heriberto marked the tree with large rocks to help us remember its exact position, as the buzzing of these bees could be more challenging to distinguish in broad daylight.

And just like that, we had searched for and found something we couldn't see. Following sounds in the pitch-black darkness, we'd attuned our senses to the symphony of the rainforest to find a needle in a haystack—bees living inside tree trunks. By walking in the footsteps of our ancestors, we'd gained a deep understanding of the mysticism and wonder these bees have brought to the Amazon for centuries. This ancestral wisdom could now guide our scientific and conservation work, ensuring that the spiritual significance of these beings remains a central part of local culture and life.

– *Stay calm.*

Roxana, Heriberto's daughter, whispered to me softly, patiently, with encouragement, as she heard me swallow in fear. Her big black eyes were the definition of reassurance.

– *Stay calm.*

I must be crazy.

The buzzing in the background was soaring so sharply, it was impossible to ignore, no matter how hard I tried. I had been biting my lips so hard they must have turned white by then, just like the rest of my face.

We had been checking out hives to find a family of bees that would be ideal for a photoshoot for *National Geographic*. Ana Elisa Sotelo, a brilliant Peruvian photographer, was here to document my work in the field. There was no predetermined story to tell, just a quest for the extraordinary.

A two-hour boat journey from the nearest town, without a fully resourced medical facility in sight, I had agreed to pose with Amazonian stingless bees on my face. Not from an agreeable distance, with protective equipment, observing them through science tools. No. On my bare face. Actually, not only had I agreed, but I had also helped conceptualize the idea alongside Ana. We'd pondered how to capture the world's attention, encouraging everyone to pause and take a closer look at these often-overlooked life forms, which are so crucial for revitalizing the Amazon through pollination, regeneration, sustainable economies, and the preservation of indigenous cultures.

In Ana's creative mind, a remarkable idea came to life—Angelina Jolie. A few years prior, Angelina had been part of an extraordinary photoshoot with *National Geographic*, where she posed with stinging bees on her face and body in an enclosed white studio. Over 60,000 specimens of *Apis mellifera* bees were attracted to the pheromone she had applied all over her body.[3] To avoid confusing the bees with multiple scents and having them think she was a flower, she'd refrained from showering for three days before the shoot.

This pheromone was the same one famously used by photographer Richard Avedon in 1981 for a portrait featuring Ronald Fischer, a beekeeper in California who posed shirtless with 120,000 honeybees.

Richard's aim was to shine light on individuals who often go ignored working in "uncelebrated jobs"—such as beekeepers. Angelina Jolie had beautifully recreated this iconic shot, drawing on her Buddhist practices to remain at ease and establish a connection with the remarkable bees. This served as a powerful platform from which to launch a new campaign promoting female beekeepers worldwide as guardians of native bees. Ana's idea was to pull an Angelina-Ronald shoot, but in the wild, with the splendor of the green life of the Amazon standing tall and powerful behind me, as I posed with stingless bees on my face.

The last hive we'd opened, we quickly discovered, was home to *Melipona titania*, considered one of the largest bee species in the world: a rarity among Amazonian native bees, which are typically smaller than the globally renowned stinging honeybee. Against my expectations, they can be surprisingly aggressive. Not in the tit-for-tat manner of most animals' understandable self-defense, but more in a "I see you there, so I will come for you" kind of way. They cling to people's hair and bodies with such determination that they can potentially tear skin apart, exposing the flesh. And they do—as I saw with my own two eyes in the split second when one of them clung to Heriberto's arm.

I couldn't stop shaking my head—no way.

Not only was my long black hair down, which gave them ample surface area to latch on to, but I also grew up with skin sensitive to temperature changes, intense sun, water quality—you name it. I went through various dermatologists and experimented with many of my grandmother's medicinal plants in our backyard—I even dabbled with Chinese Traditional Medicine while completing an internship in Beijing. So the idea of posing with bees on my face—bees that could potentially rip my skin apart, a possibility I certainly had not contemplated when agreeing to this shoot—was now twisting my stomach.

We kept searching until we found the enchanting *Melipona eburnea*, a commonly raised species that closely resembles honeybees in shape, color, and size. Known for its generally peaceful ways, my heart jumped in excitement. However, the joy quickly sank as I realized that that morning, the bees were feeling out of sync with their environment, acting aggressively, biting left and right at whomever came their way. Just my luck.

My body was frozen. My muscles had paused in spasm, certain that I couldn't endure this. I had just one day to complete this shoot, having traveled thousands of miles and borrowed time from my new university position to undertake this campaign. It was a once-in-a-lifetime opportunity for us to create a unique independent project for *National Geographic*. It was our chance to shine a spotlight on a life form that had been overlooked by the media for centuries. It felt like a make-or-break day, and there was no room for hesitation—or failure.

– You are typically calm, why are you behaving this way today? I heard Heriberto speaking to the bees. He proceeded to place some cotton inside his ears and nostrils, and asked us to give him just a few minutes alone with the bees. What I witnessed next was wild beyond imagination.

With intense concentration and visible discomfort, he closed his eyes as multiple *Melipona* bees bit him. They buzzed with such an intensity that they gave me goosebumps. The bees clung to his hair, his clothes, even his eyebrows. I flinched. Buzzing, twitching, biting. I could see Heriberto's jaw set in determination as he withstood the onslaught of thousands of bees. They could be merciless. Time, along with the restless buzzing, seemed infinite. Were we supposed to watch this unravel without doing anything to help him?

He lifted his palm as a sign for us to remain calm. One grisly minute felt never-ending. Then, suddenly, the bees began to calm down. The humming subsided, their movements became tranquil, and they began to dance around Heriberto, no longer delivering painful bites, but rather brushing their wings against him, making him laugh.

It felt like I was dreaming. Had the bees somehow recognized Heriberto's pheromones, identifying someone who kept them safe, a familiar presence that they often encountered, someone who wouldn't harm them but was, in fact, family? Had they known he was willing to endure even the most uncomfortable moments alongside them? Is this the epitome of the relationship humans could develop with nature if we tried?

– *My bees know me* – he proudly proclaimed as he removed the cotton from his head and walked away, implying it was now my turn. The bees remained calm.

Heriberto must have noticed the bewilderment, confusion, and fear on my face. What if the bees didn't take to me? What if they perceived me as a threat? What if their spirit rejected my presence? After all, I was a stranger to them, and if it was true that the bees' unusually aggressive behavior had been pacified by them recognizing and interacting with Heriberto, then my prospects didn't seem bright.

– *They know what you feel. They know what's inside. They will accept you for who you are* – he reassured me, tapping his chest. *Just breathe, and they will come to like you soon enough.*

Soon enough? My inner voice was reaching high decibels now.

Heriberto explained that we were an integral part of the forest—and if we remembered that, the bees would sense it too. There are good and evil spirits in the Amazon, and the bees possess the innate ability to discern those who mean them harm from those who don't. They help bring things to the surface and accept what arises in its fullness. They are naturally curious creatures; I couldn't fear them, for if I did, they would fear me too, and would attack.

I took a step forward.

Banishing the panicked thoughts from my mind, I continued advancing until I reached the hive. I positioned my legs firmly on the uneven ground, and without a second thought, I leaned in.

My years training as a dancer led my ribcage to move outwards and sideways rather than upwards as I tried to reach a slow breathing melody that matched the bees' humming. Thousands of bees danced around me, inspecting me, deciding who I was. I couldn't hear anything except for my own breathing and their gentle humming.

In the middle of the Amazon Rainforest, where the plants, trees, insects, and animals are so alive that you are in a symphony of music at all times, I couldn't hear anything else. The bees gracefully circled around my face, neck, and upper body, even posing on my hand to sip honey directly from my fingers, playfully offering a cheeky greeting. A feeling of a lifetime: a bee's tongue delicately caressing my skin. I must have reached such a meditative state of tranquility that one of the bees alighted on my eyelid; it was only for a fraction of a second, but it was long enough for Ana's camera to capture it.

Thirty minutes remained of the four-hour shoot. Two bees had made their way through my thick black hair and were playfully flying and humming near my white pearl earrings. A third bee tried to go up my left nostril as I wiggled my nose, squinting my eyes, trying my hardest not to sneeze. Meanwhile, another bee was busy exploring my right eyebrow, likely curious about the tiny hairs, or perhaps enjoying the texture against its fleeting wings. I understood why Heriberto was laughing with them. They were pure joy. I felt my lips curving as I tried to contain my laughter, determined not to disrupt the last minutes of the shoot.

We had applied honey to my face as a natural attractant for the stingless bees, and I felt the honey trickle down my brow, drip off my nose, and land on my lips. I made a tiny, ninja-like tongue movement to have a taste. It was like nothing I had ever had before. Thicker than water, but more fluid than honeybee honey.[4] Tasty, but not over sugary. In fact, it tasted just like some of the Amazonian fruits we'd eaten for dessert earlier that day. It was a mind-blowing delicacy that felt like silk on my skin.

The bees were finally getting very cozy. And truthfully, so was I. The adrenaline kept on giving. No water break, no sitting down break. The bees were finally in sync with me, I didn't dare to interrupt. I didn't *want* to interrupt. Just like Heriberto had predicted, I think the bees were coming to like me.

Honey holds a special place in Amazonian traditional medicine. Through ethnobotanical studies, we have learned that Amazonians use this "miracle liquid," as it's often called in the rainforest, to treat over 20 health conditions, ranging from upper respiratory conditions like asthma, bronchitis, and sore throats to ocular cataracts, inflammation, infertility, rheumatism, bruises, and kidney diseases. It's also used as a soothing remedy before sleep. In other words, it's an unmatched multitasking ingredient. During the COVID-19 pandemic, Amazonians also discovered that stingless bee honey helped alleviate various symptoms of the infection, proving to be a lifesaving remedy for many in the absence of vaccines or Western medicines.

Obsessed by the beauty encoded within nature, my colleagues, Cesar, and I became the first group to study the chemistry of Peruvian Amazon stingless bee honey, which is rich in invisible molecules with remarkable medicinal properties. It is antibacterial, anti-inflammatory, anti-tumor, and anti-cancer, and filled with antioxidants. These findings are not limited to the Amazon—similar characteristics have been observed in stingless bee honey from various parts of the world, including Mexico, Costa Rica, Indonesia, and Australia. Intriguingly, both science and history suggest that ancient honey may have played a pivotal role in our cognitive development as a human species, contributing to our evolution by providing high caloric and nutritional value that elevated our strength and brain power as a species.

Much of the medicinal power of Amazonian stingless bee honey comes from these bees' diet, although other potential influential factors are yet to be analyzed. We have observed these bees feeding on and interacting with over 40 different kinds of flowers, fruits, and trees in the rainforest, including the dragon's blood tree, the antimicrobial resin of which they harvest to construct their hives and honeypots (see page 243).

Everything about the Amazon—every river, hike, legend, plant, person, and stingless bee—has taught me resilience. Amazonians, and the life forms they coexist with and rely upon, embrace a mindful cosmovision that has endured through colonization, modernization, industrialization, and our planet's warming. They have endured some of the cruelest expressions of humankind. This alone stands as living proof of resilience and kindness, both to the explicable and inexplicable. It serves as an inspiration to accept, adapt, fight, and carry on. This is how Amazonians approach their journey with natural remedies, too—in the face of global changes, people are adapting and generating solutions that are key to their survival. Yet they are not doing it destructively. Instead, they are upholding kindness for the animals and plants that provide them with natural medicine, for balance is essential in the rainforest.

In embracing the bees' humming, their fleeting presence, the flirtatious fluttering of their wings, and the discomfort of that moment in time and space, I surrendered to nature and the fragility of my own humanity. We just needed to exist, with nothing else to do, nothing else to say, nowhere else to go. We just needed to *be*—and on that day, we simply were.

6

AYAHUASCA

Ayahuasca

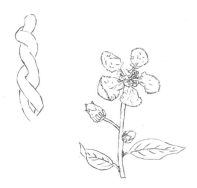

SCIENTIFIC NAME: *Banisteriopsis caapi*

TRADITIONAL NAME: Most widely recognized as ayahuasca, the brew made with this plant also goes by *yagé* (Colombia and Ecuador) and *caapi* (Brazil), among other names

ORIGIN: Native to the Amazon Basin

TRADITIONAL USES: For centuries, ayahuasca has played a sacred role in Amazonian spiritual and medicinal traditions. In Quechua, *aya* translates to "spirit" or "dead body," and *wasca* to "rope" or "vine," leading to ayahuasca being known as the "vine of the dead." Traditionally, healers crush and boil the stems of the giant *B. caapi* vine, mixed with leaves from other significant medicinal plants such as *Psychotria viridis* (known locally as "chacruna"). The specific ingredients vary among shamans and according to the purpose of the ritual, impacting the potency and the psychoactive effect of the brew. Ayahuasca's uses vary between communities and regions. It is primarily used for spiritual guidance—connecting people with expansive realms through visions and insights—and for healing emotional or mental ailments. Additionally, ayahuasca has traditionally fostered community bonding and facilitated the acquisition of knowledge about the natural world, including medicinal plants.

SCIENTIFIC INFORMATION: Extensive research into ayahuasca has revealed intriguing therapeutic benefits and complex chemical compositions. The plant *P. viridis* produces N,N-dimethyltryptamine (DMT), a potent psychedelic molecule. However, DMT is typically inactive when ingested orally due to rapid breakdown by the enzyme monoamine oxidase (MAO) in the digestive system. The vine *B. caapi* contains harmala alkaloids, which act as monoamine oxidase inhibitors (MAOIs), temporarily inhibiting this enzyme and allowing DMT to enter the circulatory system and reach the brain. Preliminary studies suggest that ayahuasca may be effective in treating conditions such as addiction, PTSD, and depression, and it may also promote neurogenesis and neuroplasticity.

– How long have you been training to be a shaman, Sergio? I asked. Twigs snapped under our feet as we passed a fallen tree with white pearl fungi decorating the trunk as thick as acrylic paint. We were leading the team back to camp after our lesson on detecting ancestral archaeological evidence of previous animal life in dried-out creeks. Along the winding paths of the rainforest, the roads seemed identical but diverged to different parts of the wilderness.

– Eleven years now . . . almost twelve – he replied, his eyes soft. A reddish-brown hummingbird zipped by so quickly its wings appeared blurry, vibrating rapidly through the mosaic of green around us. From a near distance, I sensed trees emitting a strong fragrance of garlic and lime.

– Every week, for many years, I train with the maestro *–* Sergio continued. *I have only missed it once, when the rivers flooded so badly that I couldn't navigate the* peke peke *out of my home. But I made up for it –* he added right away, smiling proudly.

As we made our way through the dense foliage, an unusually large brown and green millipede caught my eye, moving swiftly in an undulating fashion across the forest floor, followed by hundreds of leaf-cutter ants carrying leaves ten times their own size. The back of my shirt felt completely soaked from the rising humidity and the heavy backpack I wore.

Sergio was an indigenous Shipibo-Conibo man who worked as a local guide in the Boiling River area of the Peruvian Amazon. Alongside his guiding duties, he apprenticed in one of the community's most sacred traditions: the ayahuasca ceremonies.

Ayahuasca is a traditional South American plant brew that induces altered states of consciousness, and it's used in indigenous ceremonial and spiritual sessions that have been conducted for centuries. Ayahuasca is used to connect with the natural or divine world, facilitate spiritual awakening, offer personal insight, and foster emotional healing. These ceremonies are led by a master shaman, who guides participants through the ritual. As an apprentice, Sergio supported the process, learning to tune in to and navigate the wisdom of plants and animals that is thought to be conveyed through ayahuasca. The master shaman mentored him in facilitating the transmission of ancestral knowledge and protecting participants from malevolent spirits through the use of sacred songs. Training to become a shaman was less about formal instruction and more about being accompanied and guided to discover knowledge within.

Sergio openly shared his experiences with ayahuasca visions, often referred to as 'dreams' by users.

– Over the years, I've experienced many dreams... In some, I have even died, facing my greatest fears, only to return stronger and kinder.

Carefully, we controlled our pace as the path sloped downward, sand stirring up toward our eyes with each gust of wind, our boots grinding against the moving terrain of this deforested route. The absence of canopy overhead exposed us to the harsh sunlight, intensifying the heat against our skin.

– I have also seen the animals that live below – he resumed in a soft voice, looking down, avoiding the rays of sun now striking directly at us.

– Below? You mean underground, like some snakes and worms? I asked inquisitively.

– No – Sergio corrected gently, his face now carrying a veil of sadness. *I speak of the spirits of animals that once roamed the earth and now dwell in the underworld. I saw jaguars, anacondas, birds, fish . . .*

In the dreams, I learned that animal spirits remain intertwined and connected with the living. They told me they feel the anguish and pain of their counterparts' suffering due to hunting, abuse, and extinction.

– *How do you go through a dream like that, Sergio? How do you keep sane?* I whispered with admiration, imagining his shoulders carrying the weight of the world, while all the while he maintained a kind demeanor.

A playful smile adorned his face, as if he were the gatekeeper of a cherished secret. His eyes sparkled. Sergio seemed delighted by the chance to share his wisdom.

– *Calm* – he began, raising his thumb. *Control* – he added, lifting his index finger. Then, holding up his middle finger, he pronounced – *Order*.

We were now off the hill and onto flat grounds, close to reaching the borders of the community and huts.

– *As the ayahuasca medicine kicks in, the dreams will start. Some dreams can make you nervous, your head spinning, your heart beating faster. You have to listen to your own breathing, stay calm and remember the medicine is doing its work.*

I started taking mental notes.

– *The dreams may take you far away, and you may forget about your body. But your body is always present, and you are always in control. What I do is tap my fingers* – he explained, joining his thumb intermittently with his index and middle fingers, creating a tapping motion. *I tap three times to remind me I am the one in control of my body, which in turn reminds me that I am the one in control of my mind. And lastly, order. Sometimes, many dreams come at once. It's easy to get lost, and not learn everything you could learn from the plant masters. Order your dreams, so you can dive deeper into each, one at a time. There is no rush—remember that.*

– Calm, control, order – I repeated, determined to internalize these lessons as guiding compasses for whenever I felt ready for my first ayahuasca experience—and for general life. My heartbeat finally slowing from the heavy trekking, I smiled as I thought of *Eat, Pray, Love*, except I wasn't in Italy, India, or Indonesia, but rather in the middle of the rainforest, wearing cargo pants, carrying scientific equipment, and learning from ancestral Amazonian wisdom.

Studies conducted by leading experts around the world have shown that individuals consuming the ancestral ayahuasca describe retaining awareness of their surroundings. Neuroscientific and brain-imaging studies indicate that ayahuasca's effects on the brain are related to emotional processing, memory, decision-making, introspection, and visionary experiences. Some studies even suggest that DMT, the active component of the ayahuasca brew, may lead to the disinhibition of older brain systems, in a similar way to what happens when we are dreaming. In fact, reports of transformative journeys within alternate realities are not uncommon among ayahuasca users, whether they are first-timers or experienced.

For the last four days, I had adhered to the community's dietary regime in preparation for the ayahuasca ceremony. No processed foods, meats, spices, seasoning, caffeine, or alcohol; only vegetables and grains. In Amazonian shamanism, the body must be purified to access the spiritual realm.

I hadn't yet decided if I wanted to participate in the ceremony, but Sergio's words had fascinated me. My focus was on studying the ecosystem around me, yet the more deeply I integrated with the community, the more I realized that science originates from nature, and that indigenous wisdom holds an unparalleled connection to plants and animals that modern science does not typically consider. Just as the microorganisms living in this area are dependent on the soils and waters, so the act of scientific exploration is inseparable from cultural understanding.

As the sun approached the horizon, the sky transformed into a palette of colors—soft pinks, faded purples, and playful oranges that reflected off the meandering river. Darkness fell quickly, as people headed to the *maloca* carrying mats, blankets, and bottled water. The ceremony was about to begin.

The *maloca* is the ceremonial heart and cultural epicenter within the Amazonian community, constructed from wooden boards with a leafy palm roof. By day, it serves as a communal house where locals gather for social activities. By night, it transforms into a sacred space where shamans conduct ayahuasca sessions for participants and apprentices. The ceremony's leading shaman, dressed in traditional Shipibo-Conibo attire, was sitting at the *maloca*'s rear center, cross-legged, his head bowed in deep concentration. He was smoking Amazonian tobacco, and by the brief light of his pipe I could see glass jars of a chocolate-brown concoction arranged around him: the viscous ayahuasca brew that had only finished cooking a few hours ago.

I sat closest to the river, comforted by the sound of water crashing against the giant rocks. It was pitch black. Gradually, the excited chattering subsided. Over 40 people filled this expansive *maloca*, some seated, some lying down, others resting against the low wooden walls. The wind began to move, whistling as it passed through the bushes toward us, making me shiver.

We remained in silence for 30 minutes, allowing our eyes to adjust to the darkness and our ears to embrace the sounds of the Amazon awakening at night. I noticed the leading shaman, alongside his most senior apprentices, including wise Sergio, moving gracefully among the group, initiating the ceremony with Amazonian tobacco *sopladas*—an ancient ritual of blowing tobacco smoke over a participant's head and body to cleanse energy and shield from evil spirits.

The glow of the ember at the end of the tobacco pipe illuminated the shaman's face, highlighting the serenity of his gaze. Casting a small halo of light, the shaman leaned forward, exhaling a direct stream of smoke over each person's head, enveloping it in an ephemeral cloud that lingered in the air for a few seconds before dissipating into the immensity of the rainforest, mixing with the boiling-hot vapor rising from the river outside.

One by one, we were each handed a small clay cup filled with the deep-brown ayahuasca brew. It looked like thick, melted dark chocolate combined with coffee grounds. I watched as those around me confidently downed their shots, tilting their heads backward to consume it all in one go. As a cup was placed into my hands, I noticed there was more of the brew in it than I had anticipated. The smell was intense, earthy and somehow primal. I hesitated. The aroma and the daunting volume were overwhelming. I took a deep breath. And, with a gentle nod to the shaman, I closed my eyes and drank.

Wood. It tasted like wood in liquid form, warming my throat. It slowly moved down my trachea into my stomach. It felt like some of it was reaching my lungs, making me cough. Wood with a hint of dark chocolate. And leaves.

I took a deep breath as I proudly remembered my ancestral roots. My grandmother's medicinal prowess was a familiar feature of my childhood, and we often picked flowers and plants from the natural pharmacy of our backyard. We would gather and crush leaves together in our dimly lit kitchen, using large oval rocks we'd collected from nearby rivers and lakes, creating our own makeshift pestle and mortar. I remember my cousins and I turning it into a competition to see who could bring the largest, smoothest rocks in the shortest amount of time. Side by side with my grandmother, I would kneel on a kitchen chair, applying my weight and rocking from side to side, striving to release the distinct aromas of mint-like *muña*[1] or powerful Amazonian *matico*[2] before they were infused in a boiling brew. My first chemistry classes were from the wisest woman I knew.

The shaman's deep voice swept away my memories and brought us back to the lingering reality at the heart of the jungle. It began as a low humming, and then a song vibrating from his throat, growing from gentle whisper into a distinct whistle. The rhythm became more pronounced as melodic words started to weave through.

– *Pachamama... Pachamama... medicina...*

Ícaros are ancient healing songs used by *curanderos* (healers) from various ethnic groups in the Amazon Rainforest, including the Ashaninka, Shipibo-Conibo, and Quechua. These melodies, crucial to the healer's work, come to the healer during ayahuasca ceremonies, and are inherited through generations, or through "dieting" with a plant. This "dieting" involves following a rigorous regimen of consuming specific medicinal plants under a shaman's supervision to purify the body, open the mind, ward off evil spirits, and absorb the plant's wisdom. Often conducted in isolation in the jungle, practitioners may spend weeks alone while completing their diet. Some shamans I've met have even described coming face to face with the spirits of the plants they consumed, who reveal sought-after information. Some act playfully, while others are majestic and solemn, or even frightening.

Ícaros may come in many forms, ranging from whistling and singing to playing instruments like flutes. Their complexity is such that some take years to master, and the most experienced shamans can sing hundreds. When asked, shamans confess that *ícaros* often arrive in unfamiliar languages, bypassing conventional learning and spontaneously manifesting within their being without reason. These sacred songs typically feature simple phrases related to specific plants, animals, and environmental elements, each recognized for its spirit. *Ícaros* are believed to help with navigating the visions or "dreams" induced by ayahuasca, providing personal insights and fostering self-exploration—all while ensuring a connection to the physical world. It is through such vivid dreams that shamans acquire knowledge on the medicinal abilities of plants, as if spirits themselves are imparting ancient therapeutic wisdom.

At the start of an ayahuasca ceremony, *ícaros* are intended to induce a trance-like state in the participant, facilitating the absorbance of dreams.

Nothing happened for a while. I saw others yawning, which healers say is a sign that the ayahuasca medicinal brew is starting to work, as it enters your body at a systemic level. As the yawning spread across the room, some people started lying down on their mats, turning their bodies around until they found comfort.

I waited patiently for my own signs of yawning for another 30 minutes, my body feeling more awake than before. I had heard that sometimes people simply fall asleep, while others don't experience the vivid dreams that ayahuasca usually triggers, so maybe it wasn't on the cards for me tonight. Perhaps I would play a game, attempt to meditate, or just sit with my thoughts until I dozed off.

Somewhere between my impatience and meandering thoughts, I lost track of time. And that's when it started.

I felt as though someone was hugging me while holding a light behind my back. It was so bright that all I could see were glowing rays of white. I was enveloped in a deep, intimate hug, wrapped in a cocoon of uncomplicated affection and uplifting positivity. The gentle pressure was pulling me closer. It was a hug that I can only describe as warmth and security. My lungs breathed strongly, and I recognized I was still sitting down, with no one behind me—just my first dream of the Pachamama embracing me, like my grandmother does every time I return home to Perú. A sense of lasting internal peace overtook my body.

– *Are you OK?* Andrés, my dear friend and colleague, whispered. I simply smiled and nodded, lying down ready to fully immerse myself in the moment.

The senior shaman apprentices stayed sitting up, practicing their duties and mastership, overseeing everyone's experience to ensure a safe journey for all. Ayahuasca is known for provoking dizziness, diarrhea, or nausea that can lead to vomiting—a result of biochemical stimulations. I have never experienced these in my ayahuasca sessions, but these somatic symptoms are considered an important part of the traditional healing process, seen as detoxifying purges that are needed to get rid of not only toxins but also morbid thoughts and emotions. Expert healers share that if the patient is able to surrender to these unpleasant phases, a sudden and transcendental transformative experience will follow.

Sergio knelt down near me and encouraged me to talk directly to the spirit of the rainforest, loudly asking it for what I needed, surrendering to the experience, and conversing with the jungle as if with an old friend from childhood.

– Whenever you first see it or feel it, talk to it. Ask.

So I did. I asked the rainforest to talk back to me. I wanted to get to know her better, to get a little closer, as I felt I did every time I conducted science on these grounds. Having grown up between the Amazon, the Andes, and the coastline, I felt like my curiosity would never be tamed. And I hope it never is.

And as if I were rubbing my own magic lamp, the spirit of the ayahuasca delivered.

I lay with the back of my head against the wooden floor, my upper body relaxed and my knees bent. And then I heard it: the rainforest coming alive. Not in the way I was used to when traveling and adventuring through the jungle, whether by day or night. It was as if someone had tuned in to the Amazon Channel 93.0 and maxed out the amp. The sounds of every tiny life form intensified, each noise growing in sensation. Rather than being overwhelming, it was just a wave of immersive auditory experience unlike anything I had ever heard.

Nocturnal birds flapping their wings against the rustling leaves; tree frogs croaking across the vast dimensions of the Amazon; blooming flowers swaying gently with their petals brushing against each other; capybaras moving through the foliage and foraging near the water's edge; an army of ants marching over fallen leaves, creating a soft susurrus; a family of bats bursting from their hollow-tree homes and flying together in a continuous whisper. Whichever sound I focused on came forward, while the rest lingered behind as if momentarily blurring to enhance the experience, gracefully sharing the spotlight. Monkeys making their way through the treetops, jumping between lianas with agility and grace, their fingers wrapping around the vines with laser-like precision, causing them to arch slightly under their weight, producing a smack that echoed faintly in the air. Ocelots moving ghost-like throughout the thick of the forest, their sleek bodies and padded paws slipping between trees and bushes, their silence only betrayed by the crackling of twigs under their weight. Myriad insects, among them beetles, moths, leafhoppers, and mosquitoes, moving their wings at different speeds, leaping between plants, clicking, whirring, and chirping as they explored their surroundings.

And as I explored mine, I could hear the restless river, a living entity unto itself, vigorously meandering and surging with infinite energy. It collided against the jagged rocks in explosive bursts, each impact splashing water that echoed through the night air. The continuous roar blended with softer, gurgling undertones. Above this aquatic symphony, the wind swept through the *maloca*, stirring the palm fronds overhead, creating a whispering chorus and rustling percussion among the collective inhaling and exhaling of the people around me.

I felt an overwhelming sense of joy and wonder, followed by a stinging fear of not being able to hear the rainforest this vividly ever again.

And then, silence.

It was as if I had invoked my fear and turned off the radio, unscrupulously unplugging the rainforest. My eyes snapped open, my heart racing. I felt a combination of despair and dread, then recalled just in time Sergio's first and second mantras: *Keep calm and stay in control.*

I inhaled deeply and tapped my fingers three times, first fast, then slow. I felt my chest begin to relax as I regained control of my body. I could control my fingers, my muscles, my lungs, just as Sergio had said. As I reclaimed the calmness needed to proceed, a profound realization dawned on me, seemingly whispered directly into my frontal cortex: to hear the rainforest again, all I had to do was close my eyes and think of her.

The longing to reconnect with ancestral wisdom is neither unique nor new. The use of ayahuasca can be traced back 1,000 years. Archaeological exploration in a Bolivian cave revealed a leather bag containing wooden snuffing trays, a colored headband, llama-bone spatulas and traces of DMT, the natural psychedelic active molecule in ayahuasca. Similarly, chemical traces of harmine, one of the most-studied compounds within the harmala alkaloid family and the main ingredient in the ayahuasca brew, were detected in the hair of two mummies found in Azapa Valley in northern Chile. Intriguingly, this plant does not naturally grow along the Atacama coast, which indicates that there were extensive plant-trade networks reaching the Amazon. Ethnohistorical accounts and archaeological evidence also point to the prevalence of such trade networks across South America's coast and the Andes, highlighting the shared cultural values surrounding the natural world.

I wonder how indigenous communities first discovered the medicinal properties of the native plants used to prepare the ayahuasca brew. Scientifically, DMT activates serotonin receptors, which play a key role in modulating mood, perception, and cognition, acting as a short-term antidepressant. By impacting serotonin pathways, DMT alters sensorial perceptions, which may lead to the life-changing introspective experiences reported by users. Harmala alkaloids increase the amount of neurotransmitters available in the brain, assisting in emotional regulation and forming new neural connections, possibly promoting neuroplasticity. The combination of the effects of DMT and harmala alkaloids on the brain may enhance emotional empathy in the user, and could even trigger a sensation of merging with the surrounding environment, including nearby people, which may be interpreted as telepathic communication. This synergistic effect may allow participants to engage with old memories in a more insightful manner by influencing the brain's default mode network, which is responsible for self-reflective thought processes such as a memory and introspection.

Ayahuasca ceremonies are used to diagnose illnesses, treat patients, foresee future events, and connect with spirits of nature in indigenous cosmovisions. This is why ayahuasca sits at the center of Amazonian traditional medicine, with continual scientific research suggesting that this ancestral brew helps users access deep layers of the unconscious mind. Ayahuasca brings forward memories, representations, and fantasies that may not be visible to us in the conscious mind, fueling the imagination. It also enhances empathetic understanding, amplifying emotional states, earning this traditional ritual the Western nickname of "Ego Death." This refers to the personal accounts of patients who have reported that ayahuasca allowed them to confront and analyze repressed or unseen thoughts, past experiences, or aspects of themselves. Famous figures such as Will Smith and Prince Harry have described their own ayahuasca journeys in their respective books, portraying the experience as a profound insight into happiness and self-understanding, as well as an aid in processing unresolved grief, underscoring the potential for emotional healing and growth. From the Amazon to the Western world, ayahuasca is bridging the gap between deep natural wisdom and human consciousness, with

A young Rosa visiting her family in the central Peruvian Amazon with her mother.

A sign warning visitors about the hot water below.

Rosa crossing one of the hottest sections of the Boiling River.

In the Andean mountains with Rosa, her grandmother and her cousin Rodrigo as they visit family.

Catching the early-morning sunrise on a river in the southern Amazon.

The pirarucu: the largest freshwater fish in the world.

Learning from the indigenous people: the process of combing leaves to treat health issues using ancestral methods.

Rosa wearing a traditional dress with the Ashaninka people: receiving a treatment of hot vapour (*vaporeo*) infused with medicinal leaves and sacred rocks.

Rosa and her Grandmother wearing matching clothing from a local brand.

Put to the test: Rosa with the Yagua people as they assess her accuracy in blowing a dart from a distance.

A memorable photoshoot with stingless bees for *National Geographic*, with Peruvian photographer Ana Elisa Sotelo.

A poisonous caterpillar whose spikes can cause internal bleeding within hours.

In the northern Amazon, carefully crossing some logs to avoid the waters below that are home to the *shushupe* snake.

In the southern Amazon, arriving to meet an indigenous community after a long boat ride.

A pink dolphin jumping out of the water.

A scorpion under UV in the southern Amazon.

Rosa in the southern Amazon meeting with the EcoAshaninka and Reserva Comunal Ashaninka indigenous leaders to discuss the upcoming expedition.

The jaguar (*Panthera onca*), with a golden-brown coat and a pattern of circular dark spots known as rosettes.

many of its effects, processes, and experiences still waiting to be fully understood by analytical science.

Many dreams came and went throughout my five-hour ayahuasca journey. I saw ancient triangular shapes with fluorescent, vivid greens, oranges, and yellows, reminiscent of the inspiration behind many indigenous paintings and Amazonian textiles in the Amazon. I also encountered what I can only describe as the spirit of an unusually large black howler monkey roaming around, as if inspecting the space and those who came to visit. Unexpectedly, another member of our group later reported seeing a similar monkey-like entity walking around. Anecdotal reports from ayahuasca experiences indicate that some people share visions, dreams, or themes when experiencing ayahuasca within the same room. Scientific explanation or documentation in this realm is limited—most of the evidence available is based on personal accounts rather than measurable data—but indicates that the power of suggestion from a shared setting can impact the brain's perception and cognition. Shamans say that experiences like this demonstrate the power of ayahuasca, joining people in a collective level of consciousness that best mirrors the nature of the Pachamama.

Throughout the ceremony, I traversed a wide range of dreams that wouldn't fit within these pages. At some point, so many visions came that I felt compelled to call upon the third and last piece of wisdom Sergio had shared with me: order. Simply keeping in mind that I had the ability to organize my thoughts and dreams as they came allowed me to slow down time if I felt compelled to dive further and deeper into any of these dreams.

The shaman's apprenticeship involves analyzing some dreams afterwards with the leading shaman to deduce meaning and significance from them, whether in the traditional medicine realm or the emotional one. So, as good apprentices of life, my close-knit group of friends and I met the day after to share those dreams that lingered in our thoughts. I shared a particular dream that had repeated on my brain like a chorus on demand. In the dream, I had encountered words that formed a short imagist poem, vividly describing a physical space that was beyond my recognition. I wasn't able to discern its meaning. That's when Andrés, inspired by his geological knowledge and background, shared that there were only a few points along the river that fitted this description. And among those was the Spiderman Wall. This was a section of the Boiling River where you needed to rock-climb across a tall wall, with fiercely bubbling waters below you. Holding tight with your left hand, with one swift movement, you needed to push your body away, swing over to the right and grab the only distinct feature that stuck out of the rock. One push, one swing, one hold.

Regardless of the why, how, or what, this dream and conversation inspired our journey a few days later to explore the Spiderman Wall. I learned that very few had navigated it, and I became the first female to cross it. Samples sitting on the other side served as minimally contaminated specimens of microbial life in these lands.

My experience with the ancestral medicinal ayahuasca taught me that knowledge and inspiration can be found within subtle perceptions of both our inner and outer realities. This includes the language we use to express our thoughts and emotions, and to describe the environment that surrounds us. For centuries, Amazonians have engaged with nature's language in ways that are imperceptible to most, impossible to quantify, and defy conventional reasoning. Regardless of the scientific basis of these premises, welcoming the idea that nature itself can communicate has inspired centuries-long ways of living that seek to maintain harmony with the environment.

Through their indigenous worldview, Amazonians have long recognized that plants and animals convey their needs and wisdom, communicating with us if we know how to tune in. This may be through the traditional ayahuasca ceremonies by which shamans acquire deep knowledge on medicinal plants and forest resources, or it may be through profound awareness, striving for clarity of the mind and body to recognize that we can live beautifully and in sync with the spirit of our planet.

As I saw beams of light peeking through our curtains the next morning, I remembered the end of my very first ayahuasca ceremony. Earlier in the session, a deep sense of despair and a paralyzing fear had engulfed me—a fear that I had lost my ability to connect with the rainforest when all went silent. In that moment, every sound brightening my body and lifting my spirit had been abruptly shut off, the colorful orchestra of the Amazon becoming pitch darkness. For a second, it had felt like a piercing puncture in my chest. My throat had closed, and my body was struck by anxiety. That's when a primal instinct to survive had emerged, a yearning to know that the sounds hadn't disappeared. That the rainforest wasn't gone.

True to the dream's promise, all I needed to do was close my eyes and think of the rainforest. And that's when she re-emerged. The sounds and spirits of the Pachamama returned: the trees, the animals, the birds, the rivers, the insects, all as beautiful and unspoiled as ever. And along with the sounds and spirits came an infinite sense of serenity and hope: the certainty that nature was still alive and that not all was lost.

And although I didn't know it back then, it was that wisdom that would bring me back to my roots several months after my first ayahuasca experience, when the world shut down in the face of an uncontrollable and unexpected pandemic. It was a moment in time when we all confronted some of our deepest fears, including my worry that without being able to go back home, see my family, and experience the rainforest, I wouldn't be able to find peace within. I had a deep fear of disconnecting from all that I knew.

It was within desperate times of need that the ayahuasca wisdom inspired me to remember that if I ever felt disconnected from nature, all I needed to do was to close my eyes and listen intently, whether that meant tuning in to the symphony of the wild in the Amazon, following the song of a single bird amid tall buildings and grey skies, or quietening the mind to find answers within. We can all take deep breaths to gain the calm needed to step into uncertainty, knowing we are in control and can collectively build a future we are proud of. When we think about it, the spirit of the rainforest is never too far away.

7

WHAT LURKS BENEATH THE MURKY WATERS

Cat's claw

SCIENTIFIC NAME: *Uncaria tomentosa*

TRADITIONAL NAME: Cat's claw

ORIGIN: Native to the tropical rainforests of Central and South America, including the Amazon

TRADITIONAL USES: The use of cat's claw dates back 2,000 years as a sought-after plant medicine highly regarded for its multiple therapeutic benefits. It is referred to as "cat's claw" because of the unique, claw-like thorns that project from the base of its leaves. This woody vine has been used to fight infection, reduce inflammation, promote digestive health, treat cancer, alleviate arthritis, cure gastritis, and even treat snake bites. Indigenous communities use the bark and the root of cat's claw for medicinal purposes. Beyond its medicinal uses, cat's claw is considered a sacred plant among some Amazonian tribes, as it is believed to have spiritual cleansing properties.

SCIENTIFIC INFORMATION: *U. tomentosa* grows slowly in the Amazon Rainforest at elevations of 600–2,400 meters (2,000–8,000 feet), reaching over 30 meters (100 feet) in length over 20 years as it reaches maturity. Research into cat's claw supports several of its traditional uses, highlighting an avenue for its sustainable use in modern medicine. Chemical studies of the plant have detected oxindole alkaloid molecules thought to contribute to its immunity-boosting properties. Studies conducted beyond the cellular level, including animal and human research, indicate that cat's claw may stimulate the immune system and reduce inflammation, and it may also have antiviral properties, improving the patient's quality of life. This traditional plant is also being studied as a potential treatment for arthritis, cancer, and HIV. Lastly, cat's claw has been shown to inhibit the progress of local tissue damage resulting from snake bites in animal studies. Despite its multiple uses, sustainable harvesting is critical to preserving this slow-growing plant.

The first time I swam with snakes, I was 17.

My British-Peruvian high school in Perú had organized a school trip to the Southern Peruvian Amazon, and so we undertook the arduous journey into the heart of the jungle. After a two-hour small plane ride, a one-hour off-road bus journey, a two-hour boat ride down tributaries of the Amazon River, and a one-hour hike into dense rainforest, we finally arrived at the school's beautiful dark wooden lodge. It was the first time I was entering the Amazon for a reason other than visiting family or friends; we had come to do *science*. I still remember feeling important, my chest inflating with pride.

As part of our Biology and Geography course, we had been tasked to design, carry out, and complete research projects that involved working hands-on in one of the most biodiverse areas of the world: our own backyard, the Amazon. For my group's project, we had decided to study a small tributary of the Amazon River.

Soon we were waist-deep in the brown, sediment-rich waters, collecting geo-environmental data. We aimed to measure width, length, volume, and speed of the current, looking to understand the water's dynamics and any implications they might have for the surrounding riverbanks. This data could contribute to monitoring any changes in the river's health and signaling disturbances caused by climate change, such as fluctuating flooding patterns. The idea that the work we were doing, as tiny as it might have been, could have an impact when it came to protecting the river filled me with purpose.

In practice, this was a group of high-school friends putting on their rubber boots, climbing over sinking muddy hills, pulling each other out when the sinking became too intense (which makes for a great mud fight), splashing into the river, and taking measurements while laughing and messing around. It was an effortless moment of reconnecting with the beauty of the outdoors. We were pursuing knowledge joyfully, without reservation, and in doing so we learned that exploration is founded on the relationship between humans and nature. It is perhaps one of the fondest memories I have of growing up in Perú.

We repeated this process over multiple sections of the river for the purpose of collecting varied data and ensuring scientific accuracy—not for additional time playing with mud. By the time we reached the last site, however, some of us were growing impatient to return and enjoy a hearty dinner, with the water now so cold it felt like it was freezing our bones. People called out numbers and values to each other while others sat on the riverbank, jotting notes right and left. Then, someone shouted at the top of their lungs:

– *SNAKE!*

And as if time both sped up and slowed down, I turned my head just in time to see a green-and-yellow-striped snake, about 50 centimeters (19 ½ inches) long and 3 centimeters (1 inch) thick, athletically jump from the side of the riverbank and dive straight into the brown Amazonian waters. Some people gasped. Some of us remained still. But as soon as the shocked silence had built up, those of us in the water instantly shattered it—running and screaming, fighting against the muddy surface and voluminous waters to exit the river as soon as we could. Thudding, squelching, squealing.

Amid the chaos and incessant yelping, I saw the snake re-emerge from the water. With a graceful flick, it jumped through the air in a perfect, geometric oval before diving back into the river. It moved in a hypnotizing loop that reminded me of a ribbon dancer's fluid choreography as the ribbon cuts through the air. Unbothered by our screaming, the snake continued its sinuous path toward the mouth of the tributary and disappeared into the depths of these mysterious waters.

Back at the campsite, we made our way to the common room to grab some dinner. Everyone was still buzzing about the green and yellow snake that had driven us out of the river. I overheard someone recounting the story, claiming that the snake had been at least a meter long, looked angry, and seemed ready to attack us, while another insisted they had punched the snake underwater to drive it away. Amid all the fantasies and eye-rolls flying around, our teachers decided to turn on the projector and show us what we had *truly* escaped from earlier that day: a giant, muscular blackish-green anaconda (*Eunectes murinus*)[1] that our teacher had discovered while scoping out our route and diverted us away from. It was 5 meters long (about 16 feet), potentially weighing 70–150 kilograms (150–330 pounds), and wrapped around a fallen log that spanned both ends of the riverbank. It rested under the sun to thermoregulate its body, its patterned scales glistening in the sunlight, reflecting a play of light and shadow, looking magnificent. Since snakes are cold-blooded animals (also known as ectothermic), meaning they rely on external heat to regulate their body temperature, they often bask in the sun to achieve optimal temperature ranges. This aids their digestion after the consumption of a large meal, and also boosts the immune system, eases the shedding of skin, and increases muscular efficiency, making them more agile and responsive when preparing to hunt or mate.

Contrary to popular belief, anacondas are not venomous; instead, they are powerful constrictors. They use their immense strength to coil their muscular bodies around their prey, tightening their grip every time the prey breathes, constricting it until it suffocates. While anacondas may cause some internal damage or break smaller bones, asphyxiation rather than breaking the skeleton is their plan of action. They typically hunt fish, birds, and mammals, with instances of anacondas attacking humans being *extremely* rare—almost as rare as being struck by lightning. Most reports of anacondas attacking humans stem from self-defense or misidentification in the murky brown waters of the Amazon.

When threatened or cornered, anacondas can become defensive, but they are much more likely to avoid humans and will retreat if given the chance. This is why some guides recommend clapping and talking loudly while trekking through forests to deter animals from approaching. However, if it's a female anaconda protecting her eggs, her behavior may be completely unpredictable.

From the video, we couldn't determine if the anaconda was male or female. Upon closer inspection, what was evident was the absence of any visible indication that it was digesting prey, which would typically be suggested by a bulge in its stomach area due to its strong stomach acids and efficient enzymes. So, by process of elimination, the impressive anaconda we saw on screen was either boosting its immune system to potentially ward off infection, or conserving energy in preparation for its next hunt—likely a fish or a small Amazonian mammal, rather than one of the students frantically splashing around in the river.

The buzzing in the room turned into a tense whispering. We had been so close to an encounter with a giant black-and-green anaconda: us, a group of 20 loud teenagers who had screamed frantically upon running into a small, thin serpent in the wild. I smiled, feeling grateful that we hadn't disturbed the mighty anaconda or caused it any stress. We were merely visitors in its wild domain.

Curiously, my first encounter with snakes in the jungle came when I was only 12.

While on vacation, my mom and I had joined a tour in the Amazon. After an active afternoon visiting new waterways, we were back at our rainforest hotel, getting the grill ready to cook the fish we had caught earlier in the day. My mom and a few other parents were in the dining room, getting silverware and plates ready, while I played outside with some other kids. Suddenly, high-pitched screaming pierced our ears. Four moms, mine included, had left the dining room and encountered an adult cobra right outside the door. Sensing their response and fear, the cobra immediately felt threatened, raising the front part of its body, spreading its neck rib to form a hood, and hissing. The cobra had adopted a defensive posture to appear larger and more intimidating, behavior that can often result in the snake striking in defense.

As the chaos unfolded, I remember the surprising calmness I felt inside. A deep instinct took over me, as if I knew exactly what was needed. The adults needed to stop screaming, to calm down, and to slowly step back. If the cobra sensed they were not a threat and that they were willing to retreat, it would do so too. I called over to them from a safe distance, raising my voice over their screaming, asking my mom to calm everyone down and slowly go back through the dining-room door. Despite being *terrified* of snakes, my mom courageously followed through, and the fearful moment didn't escalate into a disaster. Slowly, the adults stepped back, never taking their eyes off the cobra. Once inside, the door was immediately closed, and all eyes were glued to the snake, curious about what would happen next. With the threat gone, the snake slowly retreated from its attacking pose, adopted a horizontal position on the ground, and emitted a quieter hiss as it glided away into the forest, disappearing from sight. I have no explanation for how I knew what to do. I had never faced a cobra before, nor had I seen such an encounter on TV. Yet a wisdom had emerged from deep within, overpowering fear and uncertainty, and guiding me to tune in to my surroundings, including the snake. Perhaps this is what ancestral inheritance truly means.

Many years later, these early memories came flooding back while I was journeying down lowland Amazon Basin waters during the dry season. There, I caught a glimpse of an electric eel[2] briefly surfacing, its body moving in an undulating motion that mirrored that of a snake. We were en route to visit Heriberto, the fierce indigenous Amazonian leader with whom I had previously led several scientific expeditions in the rainforest.

The low water levels had forced us to slow down to avoid the abundance of fallen trees and dense shrubbery that threatened to damage the boat's engine. This slower pace allowed us to admire the unique life forms emerging through the surface of the river. The sound of the *peke peke* quieted as we advanced more cautiously, now using paddles to navigate forward. Each paddle stroke dipped into the shallow waters, generating soft splashes.

Late into the night after it began to rain, after setting up camp and our scientific equipment, including a large microscope that the kids and I were very excited about playing with the next morning, we sat down with Heriberto's family to share a splendid dinner. Gathered around a wooden table, with myriad sounds of the living rainforest enveloping us, we exchanged stories of travels, adventures, and misfortunes. These are my favorite times in the jungle: sharing stories and dreams over a warm cooked meal. They are moments when time seems to halt, when the warmth of human curiosity transcends boundaries, allowing us to view the world around us with the wonder of a child and the wisdom of age.

That's when Heriberto shared an astonishing tale that sparked dreams beyond my wildest imagination, posing unexpected questions about life on our planet—even though I didn't realize it until many months later. It was a story that had the potential to alter our perception of animal life in the Amazon.

– *There is one more thing I fear in the deep forest,* doctora – said Heriberto, speaking directly to me.

I nearly burned my tongue with a mouthful of hot soup. I set down my spoon, already filled with curiosity. In all my years spending time with Heriberto, I had never, *ever*, heard him express fear of anything in the rainforest, aside from the *chullachaki*.

– *That day, he ran straight back home, back to our kitchen* – added Rosa, his wife, as she finished spreading the burning coal to keep the fire going. The coal crackled and popped as the steam hissed from the pot. She wiped the coal dust from her hands with a small white towel, looking at Heriberto with an expression that was difficult to read. It seemed that this memory was one of the few times she had felt truly scared for her husband's safety. He was the love of her life, the father of her children, a leader in his community, a man unmatched in his mastery of navigating the rainforest.

– *It was dry season, when the rain doesn't come and our rivers go thin, just like these months* – Heriberto said.

I appreciated his way of speaking about the rain, giving it human-like characteristics and spirit. This cosmovision is common across Andean-Amazonian cultures, where elements such as rain, wind, sun, and fire are revered as deities and gifts from the Pachamama.

Heriberto was alluding to the Amazon's two traditional seasons: dry and rainy. During the dry months, from June to September, river levels drop, rendering some waterways less navigable and making access via roads easier, thus making it a preferred time for terrestrial exploration. For the remainder of the year, water levels rise dramatically, sometimes a meter or two above the ground, flooding crops, roads, and riverbanks. Indeed, it is this prolific rain that has inspired the architecture of Amazonian homes, where the ground floor is a common space for gathering, often on the ground or cement, while wooden poles elevate the structures to tall first and second floors where families live, sleep, and seek refuge.

The dry and rainy seasons used to be more predictable. However, these seasons are now experiencing significant variations, with drought periods reaching record low water levels and extending longer than the Amazon can sustain. This is leading to increased forest fires, loss of biodiversity, dwindling water supplies, impacted agricultural practices, and endangered lives and livelihoods.

– A few years ago, my friend and I went fishing to gather food for the week. It's a bit more challenging to navigate the peke peke *in the shallow waters, but the fish tend to concentrate in smaller areas, which makes it easier for us* – Heriberto continued, breaking some bread to dip into his chicken soup. He gestured with the bread in one hand, indicating the abundance of fish they could find in a confined space, suggesting it was worth the effort of maneuvering through thin rivers. Unfortunately,

nowadays, the fish are disappearing or becoming scarce. Today, Amazonians must travel far and wide to find enough fish for just one meal.

– We had been out for a few hours already. We caught some small catfish and kept moving through the river, hoping for a larger haul. The entire trip, we were dodging tree branches and trunks right and left. In fact, we had already had to stop the engine once because shrubs and sticks had gotten stuck in the propeller, blocking the blades – Heriberto explained.

He'd had to lift the outboard motor out of the water and into the boat to clear the entangled vegetation, then carefully reattach the motor in a smooth motion to avoid any further damage.

– We were both starting to feel weary, and were contemplating whether to head back home...

Just then, Rosa gestured, asking if we'd like a *cafecito* to accompany the soup. Although we didn't need a caffeine boost to stay awake through this dinner, we all eagerly accepted, craving the rich, toasty flavor of Amazonian coffee. I got up to help with the cups, but she tapped me on the back and shook her head, signaling for me to sit back down.

– Suddenly, we hit a large tree trunk – Heriberto went on, his eyes wide open, completely immersed in sharing this surreal experience with his friends. *The impact was so intense we almost fell off the boat.*

He mimicked his body's reaction to being jolted wide awake by immediate danger.

– The tree trunk, which was floating just below the surface of the water, didn't budge. We had been bumping into so many branches in these dark brown waters, I was already feeling impatient and annoyed, so I got out of the boat to move the trunk and be on our way, you know?

I nodded, thinking back to our journey earlier in the day, zigzagging through plant matter in the shallow waters.

— *Doctora, I've moved hundreds of fallen tree trunks in my lifetime. Maybe even thousands. I know how it works. I do the exact same thing every time.*

Heriberto paused, as if contemplating his next words. An unexpected voice inside my head predicted what he was about to say, but . . . No, it couldn't be.

The rain outside had now slowed, turning into splashes against the ground, accompanied by a soft wind weaving through. The sound of a few chickens clucking and cooing gently floated in from outside the house. The tiny, beautiful family kitty emerged from a corner where it had been hiding and headed under the table, purring, its fluffy gray tail brushing against everyone's feet.

Heriberto stood up to better illustrate what had happened next. He indicated that when he'd stepped out the boat and into the river, the water had only reached up to his chest.

—*So I leaned down and put my arms around the trunk. Except that . . .* Here, he bent his arms as if hugging an invisible ball, unable to touch his fingers by a few inches. *I couldn't fully grasp it; it was really thick. The waters were so turbid and filled with plant litter, I couldn't really see beyond my two hands.*

Rosa closed her eyes and turned her head to the side, as if she didn't want to hear what had happened next.

— *I put all my weight backwards to pull the trunk out of our way. I tried and tried, but it wouldn't move a centimeter. It was so heavy and felt like smooth wood.* He changed his body movement now, pushing his arms forward. *So I tried pushing it instead, as hard as I could, and that's when I heard a scream.*

Heriberto's friend had yelled as loudly as he could, urging him to jump back into the boat. In a fraction of a second, Heriberto had sensed the terror in his voice, seen the way his face had turned white as though he had just seen a ghost.

Or worse.

The head of a giant snake.

With his heart in his throat and his lips dry, Heriberto pulled himself back into the boat, moving as quickly as he could. The boat tilted to one side under his weight, causing him to stumble as he reached for his paddle. His friend had already started paddling frantically, trying to move away as fast as possible.

Heriberto dared to glance back one more time. A few meters downstream, what appeared to be the top of a giant snake's head dove back into the water, causing a splash. Its body barely moved, yet it created a wave noticeable enough to give them an advantage by creating a ripple effect that pushed the boat slightly further away. They couldn't see the tail, nor gauge the full length of its body. All they knew was that had Heriberto grabbed the snake a few meters up, he would have encountered the head of a predator so incredibly powerful it could have ingested him in one bite without any effort at all.

They didn't utter a word for the remainder of the trip, both paddling as strongly and quickly as they could. It was only upon their return to their village, after they had parked the *peke peke* safely and checked their surroundings to ensure the murky waters were clear, that Heriberto and his friend had hugged for an extended moment, thankful that the rainforest had spared their lives.

The campsite sat stunned at Heriberto's story. I had barely blinked, trying to catch every detail of his gestures and expressions, my eyes feeling dry from the intense concentration. My mind raced with questions, looking to connect the dots and make sense of this tale. Heriberto's words had challenged everything I knew about wildlife in the rainforest.

For a moment, I decided to focus solely on the facts. First, it is well known that the green anaconda of the Amazon Rainforest is the world's largest and thickest snake. However, the largest documented anaconda measures about 30 centimeters (12 inches) in diameter—roughly the size of a dinner plate—not nearly as thick as a tree trunk that you can barely hug with both arms. Second, water levels in the rivers have reached record lows due to drought, exposing life forms and geological structures that usually lurk beneath the murky waters. Third, much like the ocean depths, which are home to creatures still unknown to mankind, many regions of the Amazon remain unexplored. Every year, scientists uncover new marine species at the bottom of the ocean, highlighting how little we know about the fauna that resides at far depths.

So, could it be remotely possible that there was an unexplored fauna living at the bottom of the Amazon River and its tributaries?

Until then, my sole focus had been on documenting and studying wildlife on land, not in the murky brown waters of the river. But Heriberto's story had shattered any preconceived ideas I'd had about the aquatic life of the Amazon, instilling a deep curiosity and a desire to explore further.

– *You are asking about the* Sachamama – said Avi in a calm voice. Avi was a remarkable biologist and extraordinary local guide with vast knowledge of the virgin rainforest. I had reached out to her looking for clues about the existence of a record-breaking giant anaconda.

– *The* Sachamama? I repeated, recognizing the Quechua words for "forest" and "mother."

– *The legend started in the time of* los antiguos *(the ancient ones). My grandfather told my dad that an Amazonian hunter once went into the deep jungle looking for prey to feed his family. However, he wasn't getting any luck. It started raining heavily and he heard animals screaming. Confused and terrified, he knew it would be unsafe to move and waited until the rain cleared.*

– *The next morning, the hunter found a giant hole in the ground, so large that it resembled a cliff. He descended into it, noticing roots intertwining everywhere. Once deep in the hole, on the forest floor, he turned on his light...* Here, Avi paused to drink a sip of water, then went on... *and found a giant snake.*

– *A giant snake underground?* I asked, completely immersed in her storytelling.

– *Yes, the legend says that the* Sachamama *was so big that she lived below the ground, under the trees, with roots adapting and growing around her body. She needed a lot of food to survive, and so she kept her mouth open, pulling surrounding animals into the hole, never to be seen again.*

According to legend, the *Sachamama* possesses a powerful allure, capable of hypnotizing anyone who locks eyes with her, compelling them to venture deep into the rainforest to become her nourishment.

I remained silent for a few seconds, processing Avi's words and the extraordinary details of the legend. And that's when she shared another important clue.

– *My grandfather said this happened around kilometer seventy or eighty, before it was turned into a road.*

I rummaged through my bag, reaching for my notes, skeptical of the idea that Heriberto and Avi's stories would overlap. They didn't know each other. There was no way. Except that there was. The specific area that Avi was describing was within the parameters of Heriberto's territory. I felt my heart palpitating as I discovered the connection.

– He said the military showed up shortly after word of the hunter encountering a giant snake went around. They went into the deep forest and killed the snake. My grandfather remembered the helicopter flying a few times back and forth over their heads, with elders saying that the snake was so big the military had to cut the animal into pieces, carrying one body part at a time.

The *Sachamama* is a deity that is frequently depicted as a giant snake dwelling deep underground in the Amazon, in a state of lethargy. She is revered as a symbol of wisdom and fertility, believed to influence the growth and health of flora and vegetation, and to hold the power to communicate with other spirits of the forest. Amazonians regard the *Sachamama* as an ancient guardian that protects the rainforest and all its inhabitants.

The legend of the *Sachamama* is often told to children as a moral lesson, emphasizing the significance of the rainforest and the imperative to protect and preserve it for future generations. This deity also serves as a cautionary symbol of the Amazon's dangers, reminding all of the necessity for caution and respect when navigating these vast, untamed landscapes.

Over the years, stories about unusually large anacondas or snakes lurking in deep holes beneath giant trees, waiting for prey to stumble in, have surfaced throughout the Amazon, from the North to the South. Similarly, tales of giant anacondas measuring 12–15 meters long (40–50 feet) have been told across the tropics worldwide, although never verified. Through my travels and conversations with family,

I've encountered these stories as well, but always indirectly, through a friend of a friend, always just distant enough to warrant deeper questioning.

Thus, neither Heriberto nor I had ever thought twice about the idea of a giant snake roaming in the Amazon beyond the legend, the ancient depiction of the *Sachamama*, a product of human imagination.

Until now.

Whether these local tales blend myth and reality, one fact remains undeniable: snakes that dig and live underground are real. They are known as fossorial snakes, and some are, indeed, found in the Amazon Rainforest. These reptiles have adapted evolutionarily to spend most or all of their lives beneath the surface, evading predators and preying on subterraneous food such as insects, worms, and other invertebrates. While some fossorial snakes primarily live underground, they may occasionally exhibit terrestrial behavior.

Fossorial snakes vary in size, diet, and behavior, with many possessing a head shape conducive to burrowing. Their vision is often diminished or non-existent, as eyesight is less necessary when living in perpetual darkness. Instead, these snakes have highly developed senses of smell and vibration detection, contributing to their secretive nature. Although it is hard to assess how many fossorial snakes we have in the Amazon, it's known that deforestation and urbanization disrupt and compact soil, potentially restricting their underground movement, sometimes forcing these creatures to surface and endangering their existence.

Currently, the Amazonian green anaconda holds the title for the largest snake on our planet, including those with darker colorations, like the one we nearly encountered during our high-school trip to the rainforest. Some reports suggest these snakes can reach lengths exceeding 8 meters (26 feet), with females significantly larger than males.

However, the green anaconda becomes minuscule when compared to snakes that existed in the past, like the *Titanoboa*. This prehistoric serpent, known as the largest snake to ever inhabit our planet, lived approximately 60 million years ago during the Paleocene epoch, shortly after dinosaurs went extinct. Fossils of this behemoth were discovered in northeast Colombia, not far from the Amazon Basin. Research indicates that the *Titanoboa* thrived in a warm and wet environment akin to today's Amazon. Believed to have been a semi-aquatic predator, estimates suggest this ancient reptile could grow up to 15 meters (50 feet) in length and weigh as much as 1,100 kilograms (2,500 pounds), a size facilitated by the warmer conditions of the time.

The discovery of the *Titanoboa* has opened a window into understanding the post-dinosaur extinction era, marking it as a pivotal species in the evolutionary narrative of life on Earth. As the planet cooled, the size of cold-blooded animals became limited by metabolic constraints. Following *Titanoboa*, remains of other large prehistoric snakes, albeit smaller, have been found in North Africa and Australia, with the most recent specimens discovered in Australia dating back approximately 50,000 years.

In 1959, Colonel Remy Van Lierde, serving in the Belgian Air Force, reported encountering a giant snake while flying over the Congo in Africa. During this flight, he and his team observed a massive unidentified snake on the ground. As the helicopter descended to take a closer photograph, the snake reportedly reared up, as if preparing for an attack. Charles Hapgood, a Harvard alumnus and history professor, noted that there had been "countless reports from the . . . area of giant snakes," and stated that verifying the authenticity of the photograph would "be a great triumph."

Experts and zoologists examined the old black-and-white photo, comparing the snake's size against nearby termite nests and trees, as well as estimating the helicopter's height, to suggest that the snake measured 12–14 meters long (39–46 feet). Unfortunately, further analyses of the photo were not conducted, leaving additional verification unavailable. This report remains one of the most intriguing pieces of evidence in the ongoing quest to document the existence of modern giant snakes, yet it also highlights the challenges in validating such extraordinary claims.

Keeping an open mind to stories that challenge the ordinary does not devalue our expertise and intelligence. In fact, it can inspire conversations that reveal answers to questions we'd never thought to ask. That's exactly what happened to me a few months after hearing Heriberto's account.

Back in Lima, my family and I were gathered for Sunday lunch at our home. I had recently returned from England, and was still feeling the jetlag of intercontinental travel. Everyone gathered around the table, with the bustling sounds of a busy city outside: cars honking, sirens wailing, people chatting. Our wide-open windows invited in the fragrant breeze from the natural pharmacy my grandmother continues to nurture in our backyard, made up of Amazonian and Andean bushes, trees, plants, and flowers in soil and carefully arranged pots, each tagged with its scientific and traditional names to honor the heritage of our home—a sacred blend of indigenous wisdom and modern science. They serve as both a culinary repository of herbs and spices, and a medicinal trove for curing all that needs healing, from coughs, infections, and stomachaches to headaches, anxiety, and low moods.

We were eating *lomo saltado*, a quintessential Peruvian dish consisting of beef sautéed with onions and tomatoes, served with delicious white rice and fries dipped in oil. There is an old saying that goes: "*Los peruanos tenemos derecho a comer rico*," which translates to "We Peruvians have the right to eat well." I think this demonstrates just how seriously we take our food. It's deeply embedded in our culture that meals are sacred times for reconnecting with loved ones and passing down generational knowledge. It also means we have delicious recipes to suit all tastes and desires—all the time.

The atmosphere was vibrant, and everyone excited to be reunited. As I offered more rice for refills, I was recounting my recent travels, eventually delving into Heriberto's unforgettable story, which still lingered in my thoughts. My grandmother made a face and shivered at the retelling; she has never been a fan of snakes. My curious younger cousins wondered aloud whether I'd ever brave an expedition to search for this colossal reptile. Repeating Heriberto's gesture, my mom stretched out her arms in a wide circle to illustrate the significant thickness of the snake that Heriberto had mistaken for a fallen tree in the Amazon's murky waters.

– *But Rosita, you know those are real, right?* my usually reserved uncle interjected, capturing everyone's attention. Starting in the 1970s, he'd spent nearly two decades in the Central Amazon working as a flight controller at a local airport. The room fell silent.

– *What do you mean?* I asked, as I set down the rice and began serving some icy homemade lemonade. Was I about to hear another first-hand account of the possible existence of this enigmatic creature?

– My colleague at work was a night fisherman – my uncle began. *He used to go out in the river alone, or sometimes with a friend, in the middle of the night, when everything was quiet and pitch dark, as he said he always caught his largest fish at that time.*

My uncle described his colleague as a hard-working and earnest man, always honest and genuine: a good friend with whom he'd shared many beers.

– One night – my uncle went on, leaning in, *– he was sitting at the rear of the boat, his friend opposite him to balance the weight. His friend was closing his eyes, taking little naps from time to time. Then, without any warning, a giant anaconda emerged.*

As he spoke, my uncle's hands traced the path of the snake in the air to illustrate its movement.

– It slithered into the boat from one side, rocking them sideways, and then slowly crossed the bottom of the boat, zigzagging its way through.

My grandmother shivered again and stood up to head for the kitchen with the excuse of getting more fries.

My uncle elaborated on the sheer terror his friends had experienced. Frightened, they'd stared at each other, becoming paralyzed due to a primitive instinct to survive. They tried holding their breath, terrified that the snake would hear them and come straight at them, right on time for its midnight snack. The anaconda's girth was massive, suggesting it could easily swallow an adult man with a single bite. The machete, their only potential weapon, lay out of reach, and even if they could have grasped it, piercing the snake's thick skin would likely have provoked a deadly confrontation, ending with them in the water, where they would stand no chance in the complete darkness.

They held their breath until they were dizzy, while the snake made its way across the boat, seemingly either unaware of or indifferent to the humans visiting its waters. Eventually, as effortlessly as it had appeared, the snake slid back into the dark waters, vanishing from sight. The men finally exhaled, their pulses thrumming with adrenaline.

Looking up, wondering what I might decide to do with this information, my cousin Rodrigo spoke softly from across the table.

– *Another giant snake...*

Despite inspiring fear, Amazonian snakes have been a great source of inspiration for new, lifesaving, groundbreaking medicines. If you have ever lived with anyone with unusually high blood pressure, they might have been prescribed a medication known as "captopril." I first learned about this pharmaceutical because my dad took it for years. Intriguingly, workers in Brazilian Amazonian banana plantations had been seen fainting unexpectedly due to sudden drops in blood pressure. Quickly, locals learned this occurred when the men were bitten by a local *Bothrops* snake, a pit viper whose venom contains peptides that inhibit enzymes responsible for raising blood pressure, effectively causing vasodilation and lowering blood pressure. This observation led to the development of synthetic molecules that mimic the venom's action, resulting in the creation of captopril in the late 70s. Since then, similar drugs have been introduced to the market, benefiting patients with heart failure and hypertension. Thank you, Amazonian snake!

Parallel to these scientific discoveries, native plants in the Amazon have long been used to treat snake bites, such as the ones that injured the men working on the banana plantations. An example is cat's claw (see page 140). My grandmother used to give us cat's claw tea to boost our immune systems and energize us when we were recovering from the flu or any common illness.

Ever since, the experiences shared by Heriberto, Avi, and my uncle have lingered in my mind. These tales challenge my understanding of the world's mega-fauna and raise the possibilities of giant creatures surviving past extinction events that eradicated most of their kind in tropical regions like the Amazon. These stories open up an infinite pool of questions about what else might exist in the unexplored depths of our world. They defy our current knowledge, encouraging us to be brave enough to reconsider the mysteries that are waiting to be explored.

That's the essence of exploration and storytelling: the sharing of human encounters with nature to broaden our perceptions of what's possible. It's not always about the act of discovery but the pursuit of knowledge. Sometimes, it's about introducing an unexpected dimension of beauty and mystery to our planet, breaking old patterns of thinking, and urging us to remain inquisitive. It encourages us to continue exploring, and to remain open to the endless possibilities that can shed light on the complexities of our natural world. Perhaps it is by examining the Earth's ancient epochs that we can devise innovative ways to sustain and regenerate life forms and ecosystems in the present.

8

LIVING FOSSILS

Copaiba

SCIENTIFIC NAME: *Copaifera* spp.

TRADITIONAL NAME: Copaiba

ORIGIN: Native to the Amazon Rainforest

TRADITIONAL USES: Copaiba is a highly valued natural resource for Amazonians, particularly in folk medicine. The tree exudes a sticky and aromatic resin or oil when tapped (much like one would tap a maple tree for syrup). This oil is renowned for treating skin conditions, as it promotes wound-healing and reduces inflammation in cuts and bruises. It is also often used as an analgesic, offering relief from muscle pain and arthritis. The resin is also ingested to combat urinary infections and aid digestion. Beyond its medicinal properties, copaiba oil is an effective insect repellant and doubles as a delightful natural perfume, boasting unique fragrance tones.

SCIENTIFIC INFORMATION: The copaiba tree, belonging to the genus *Copaifera*, grows to heights of 15–30 meters (50–100 feet) and is distinguished by its small, fragrant yellow or green flowers, along with small, round red fruits that contain seeds. Though not among the Amazon's largest trees, the copaiba stands out as a large mid-canopy tree, making it a prominent figure in the jungle and highlighting the need for sustainable harvesting. It reaches sunlight effectively and also plays a significant role in enhancing the forest's biodiversity. Copaiba oil is rich in bioactive molecules, including sesquiterpenes and diterpenes, which bolster the oil's capacity to fight inflammation and foster skin regeneration. A key component is beta-caryophyllene, a cannabinoid receptor antagonist known for having analgesic and anti-inflammatory properties without inducing psychoactive effects.

Legend says that once there was an Amazonian warrior named Pirarucu who was known for his incredible strength, but also his excessive vanity and selfishness. According to myth, this man exploited his undeniable power, disrespecting nature and his fellow humans, taking community members hostage, and annihilating them without reason. Tired of his behavior, the gods decided to punish him while he was fishing on the riverbank without the community's consent. The gods unleashed lightning, thunder, and torrential rains that caused Pirarucu to fall into the water—and then they cursed him, causing him to transform into a gigantic scaly fish destined to live forever in the Amazon's waters.[1]

The pirarucu continues living in the deep Amazonian waters, now representing the largest freshwater fish of our planet; a fish so giant that it can reach the size of a classic Volkswagen Beetle. Besides its impressive size and cultural legend, the pirarucu also serves as a window into the ancestral aquatic life of the Amazon. Known as a "living fossil," this species has undergone minimal evolutionary change from its ancestors. This means that its genomic profile, ecological niche, and physical characteristics have remained virtually unchanged for millions of years. The pirarucu is a testament to a world that has long vanished from the surface of the Earth, yet still whispers through the wildlife of the jungle.

—*STEP BACK!*

I heard our Amazonian guide Rufilio yell across the gazebo built over the side of the lake where a family of pirarucus lived.

An orange-striped cat stood startled with a stiffened tail. It had strayed perilously close to the water's edge, its curiosity betraying its position. The cat elegantly retreated after this warning from its owner, narrowly evading the possibility of becoming a midday snack for the giant lurking in the water.

I'd arrived just in time to witness the upper portion of the pirarucu's body vanish beneath the surface. Its dark green and brown scales shimmered like jewels against the turbid waters and the green ovate-leafed plants floating on the lake's surface. The two black eyes of the fish, comically set apart, reflected upwards through the water. Its large, upturned mouth, bigger than my head, seemed to flash us a cheeky smile before it dove back into the depths, its dorsal fin stretching toward its immense red tail. Unlike most fish, pirarucus are obligate air-breathers, meaning they are required to surface for air about every 20 minutes, emitting a loud, coughing sound. The fish has a specialized swim bladder connected to its mouth, acting similarly to a lung. Kneeling, I squinted, attempting to trace its path—and spotted three other pirarucus blending seamlessly with the murky water and the green reflections of the surrounding foliage.

I was in the northern part of the Amazon with my team to learn more about these living fossils at a sustainable pirarucu farm. Pirarucus are not known to be aggressive toward humans, but, should a person accidentally fall into the Amazonian waters, there's a risk of being inadvertently struck by one of their tails—which, considering their 200-kilogram (440-pound) mass, could potentially cause significant injury. It was a thought that crossed my mind as I slipped on a watery patch of the wooden platform that overlooked the three giant pirarucus before quickly recovering my balance.

Laughing and carrying a small purple bucket brimming with tiny fish, Rufilio approached me and Chris, wielding a pole with a fish dangling from its end. As he extended the pole over the water, two of the pirarucus leaped agilely, each mouth gaping to create a vacuum. They inhaled their snack, sending water flying. Despite their hefty, muscular bodies, these fish are swift swimmers and adept hunters, capable of using bursts of speed to leap from the water and snatch low-flying birds, lizards, and even small mammals that venture too close to the water's edge.

– *Their skin and scales are so thick that not even piranhas can bite through* – Rufilio told us with admiration in his voice.

We observed the pirarucus meandering under the water, waiting for more food.

– *Their scales are great for making shoes, purses, bags, and art . . . like your earrings!* he continued, noticing my pair of long blue earrings, a beautiful gift I'd received from an Amazonian woman a few years back.

Traditionally, Amazonians also repurpose pirarucu scales as nail files or to craft spoons and spatulas for the kitchen. Even their tongues can be transformed into tools for grating yucca or used as sandpaper to smooth wood.

– *Can you believe that its cousins lived around the time of the dinosaurs?* Rufilio asked, playfully mimicking a dinosaur's tiny arms with a chuckle, before tossing more fish into the lake.

Cue splashing, gulping, jumping.

– *Pirarucu, valiant and stubborn warrior* – he murmured, continuing to feed the fish. As the giant pirarucus snatched the food, we were showered with lake water. I closed my eyes for a second, enjoying the cool sensation of cold water splashing on my skin amid the oppressive heat, right as a family of macaw parrots flew by, emitting loud, high-pitched screeches and casting colorful reflections of blue, red, and green on the brown waters.

*

The name "pirarucu" derives from the Brazilian Tupi language, roughly translating to "red fish." Generally known as *Arapaima gigas* in scientific terms and colloquially called "paiche" or "arapaima," this creature has been recorded as reaching lengths of up to 4.5 meters (15 feet) when adult, demonstrating the fastest-known growth rate of any fish worldwide. Long thought to be a single-species animal, scientific research has revealed that there may be five or more species of pirarucus in the Amazon.

The pirarucu is considered "the cod of the Amazon," and is often regarded as a cultural symbol of fertility and abundance, representing a vital food source for Amazonians. The practice of salting and drying the fish's meat allows for long-term storage in an area largely devoid of refrigeration. Given its considerable size, a single pirarucu can feed numerous family or community members, yet its lean, protein-rich meat has also made it a coveted delicacy, spurring overfishing and endangering its survival. In response, conservation efforts and sustainable fishing practices are being employed to cultivate pirarucu, aiming to bolster economic growth without compromising wild populations. The farm we were visiting raises a limited number of pirarucus annually, each kilo selling for approximately $30 USD, thus offering a dependable revenue stream for local families.

The pirarucu is considered one of the oldest fish species in the region, boasting a genetic lineage that stretches back millions of years, to a time just after the Jurassic period. The pirarucu has remained virtually unchanged for a vast span, representing a direct link to the Amazon's past. This fish reveals the biodiversity that once flourished under climatic conditions predating human existence. It represents the resilience of life, demonstrating that some species did not go extinct, but instead adapted in harmony with their surroundings, including us.

We retraced our steps across the slender planks, leaving the sanctuary of the gazebo behind. Hundreds of tall trees rose around us in the family's property as I rubbed copaiba oil on my arms to fend off the mosquitoes that were starting to feast on my skin. Noticing my small oil bottle, Rufilio gestured toward a towering and majestic copaiba tree in the distance. Behind it rose tall huts on wooden stilts, the elevated homes where various family members live, each marked by small, unique symbols as if to name each home.

I looked around to take in the fresh air; I love visiting family-owned sustainable farms in the Amazon, whether they be for stingless bees or pirarucus. The enduring wisdom of Amazonians shines through as they cohabit peacefully with nature, building a circular economy that not only safeguards but also celebrates biodiversity, land, and water; they offer a haven where life thrives, shielded from the threat of rampant deforestation. Amazonians carry valuable lessons on integrating the cultural importance of nature and life forms with ecological wisdom and community impact, achieving a holistic approach to conservation.

Before heading out to our next stop, we took some time to simply enjoy the beauty around us. Surrounding the natural sanctuary was a vibrant array of aromatic flowers, orchids, and bushes, including *Heliconias*—known as "lobster-claws"—which are prevalent in the rainforest, with green, orange, yellow, and red bracts attracting hummingbirds right and left. I also spotted the "monkey brush"

(*Combretum rotundifolium*), a tropical plant decorated with bright yellow and red flowers that stand out in the green abundance of the Amazon. I hadn't seen any in quite some time.

– *Holaaaaaa!*

The acute, piercing call suddenly echoed around us. The dense trees obscured my view, so I couldn't see who was calling, but Rufilio, privy to something I was unaware of, burst into laughter and quickened his pace toward a fork in the wooden path. As I upped my speed following him, I heard the voice again, except now it sounded like multiple voices, simultaneously greeting us in a high-pitched tone.

– *Holaaaaa* – I called back, only to discover a family of three green-and-yellow parrots, delighted to find that they were able to summon visitors with their calls.

– *Helloooo, hiii.*

– *Did they just speak in English?* Chris inquired, astounded, responding in English himself.

We too started laughing, with Rufilio still cracking up a few feet away, looking at the scene, always amused by the parrots' knack for drawing people to their gathering spot. Chris had not heard much English spoken on this trip, so I could see him enjoying the communication—even if it was with a parrot.

The sanctuary was a mirror of the parrots' wild home, with bowls of water and a variety of the seeds and fruits beloved by the birds, all strategically placed to minimize competition and promote the well-being of these colorful and playful creatures. On one side, a large whiteboard detailed their names and their rescue story, revealing how they had been cruelly stolen from the rainforest and taken into cramped city apartments, where they were chained by the feet. The constant friction of the chains rubbed them raw, and the signs of their pain were still visible.

– Hey gentleman!

The parrots happily continued the conversation, shaking and bursting into laughter. My own chuckling was now uncontrollable. The family's children had now joined us, standing beside Rufilio, covering their mouths and clutching their bellies.

– The voice of the jungle – Rufilio announced, referring to the parrots' vocal mimicry.[2]

A moment of pure, simple happiness—a vivid reminder that in the Amazon, the bond with animals transcends mere respect; it is also a source of joy. We'd felt that joy in the cool splashes of water on our faces as the pirarucu leaped for its meals, and now we heard it in the vibrant laughter and voices of parrots. The profound indigenous appreciation for these creatures fosters a strong sense of camaraderie, motivating local families to protect the biodiversity of their home by courageously combatting the human threats and climate change impacting wildlife.

We set off in the boat, each of us still wearing a large smile as we navigated toward our next destination of the day: a family-owned sustainable monkey sanctuary.

The sky was filled with strokes of pink and yellow, large clouds adorning the green backdrop of everlasting Amazonia. The breeze felt like silk against the skin. My hair was loose, flying behind me, my neck cooling down as droplets of water leaped into the boat.

After a turbulent 30-minute journey along the river, we finally arrived at the sanctuary. This haven offered a peaceful retreat for rescued and abused monkeys to recuperate, gradually reacclimatizing them to the wilds of the rainforests, though some opted to stay within the sanctuary. We were here to understand more about the strategies utilized by the local family to facilitate the animals' recovery and re-adaptation to the wild.

Maria and José, the sanctuary's founders, welcomed us warmly. This devoted couple have poured their lives into restoring a swath of the Amazon and rescuing monkeys from the cruel fate of illegal and frivolous entertainment. They demonstrate first-hand how a harmonious coexistence with the Pachamama is possible—and essential.

Their space was vast, extending a few hectares into the deep jungle, with tall trees embracing a peaceful home in which the monkeys could recover their health. Sprinkled throughout the reserve were bowls of clean water and fruits for the animals to enjoy. Their office, a humble wooden hut, stood at the right-hand corner of the campus, offering a space where visitors could rest from the sun and learn more about their rescue methods by viewing photos and posters.

In the center, a poma rosa tree (*Syzygium* sp.) stood majestically, towering meters above us, its thousands of striking bright pink flowers resembling a giant stick of cotton candy. The flowers were both atop the tree and scattered on the floor as they were shed to make room for glossy, juicy, deep-red fruits that had a crisp texture and were mildly sweet. The immensity and beauty of this tree is beyond belief.

The endless palette of colors that I find in the Amazon makes me feel safe. From when I was a little girl, when my grandmother would teach me about the flowers that bloomed in our garden, I remember eagerly searching my Faber-Castell colored pencils for tones that matched those blossoms. I would spend hours combining colors and playing with techniques to achieve the tones and textures that I would find in the flowers. Velvety, silky, papery, prickly. Pinks, oranges, greens, blues. I was always dreaming of color. Our garden was a rainbow of life. This obsession for decoding the color of nature only grew when I learned how to use a field microscope to reveal the hidden hues and textures of the wild. The pursuit of color in nature powers human curiosity and awakens our inner child, encouraging us to revel in the joy of beauty in the outdoors.

I ran below that tree as if I was five, playing with the fluffy, soft, feather-like fallen flowers, composed of numerous delicate filaments. I gathered handfuls of them, bringing them closer to my face to savor their gentle yet sweet fragrance, which draws hundreds of pollinators daily.

One of the family's youngest children emerged from behind a tree, his eyes wide with wonder and a smile illuminating his sun-kissed cheeks. In his arms sat a young sloth that clung gently to his shoulders. The white patches around its eyes contrasted with its dark mottled fur, resembling a superhero mask. Its rounded dark nose gleamed in the sunlight. The sloth wore a perpetual smile, its half-closed eyes fixated on me with a serene expression. The deep-brown beautiful lines marking its small, round face radiated sheer tenderness as the kid stroked its back.

– *Mi amigo* – the kid murmured, drawing closer to introduce me to his companion.

The boy reminded me of a distant cousin in the Amazon who, in his youth, used to play with monkeys and sloths in his backyard, teaching me the gentle art of interacting with animals. One never dominated, imposed or took the animal in as a pet; instead, the interaction was a cultural practice among Amazonian families that must not be commercialized or encouraged with visitors. They always acted in harmony and reciprocity with nature.

I waved as the sloth reacted with measured curiosity, its movements slow, almost imperceptible at first. It lifted its head, completely unhurried, and its eyes gradually widened, revealing tranquil awareness, while its rounded ears twitched softly, attentive to any sounds around. At a snail's pace, it extended one of its velvety limbs, looking to explore the air between us. It was a slow and subtle gesture that indicated it felt comfortable in my space. Another living fossil.

Fascinated, I returned the sloth's gaze with a big smile. As I looked at the small creature before me, I tried picturing the size of its ancestors.

Few people know about ground sloths, the real *giant* sloths, the size of four adult men standing on top of each other, that once roamed free from Alaska to Argentina. There is no debate among paleontology experts that these colossal, fascinating mega-fauna existed in our world, from North to South America. As if that's not impressive enough, it appears that some of these ground sloths persisted beyond the most recent major extinction phase in our planet's history, known as the Quaternary extinction event or the last Ice Age. This period marked the end of many large mammals and other species that once occupied our lands and oceans, such as mammoths, mastodons, saber-toothed cats, glyptodons, and more.

Remnants of *Xenarthra* ground sloths were found in the West Indian islands, among the last regions in the Americas to be settled by humans. Not native to the islands, it is thought that the sloths arrived there by swimming or floating from the mainland, and that they were able to survive a few thousand years longer than their relatives due to their isolation from early human hunters, going extinct only about 4,400 years ago, a timeframe that represents barely a second past midnight on the clock of evolution.

Among the largest mammals to ever walk our planet, the ground sloth, from the genus *Megatherium*, ranks top of the line, as it could grow up to 6 meters (20 feet) tall and weigh as much as 4 tons, resembling a modern elephant. They were not endemic or confined to one specific area. Instead, they thrived prolifically across various ecosystems, from the swamp that the Amazon Basin was before becoming the Amazon River to the Andes Mountains, back when the mountains stood at only a quarter of their current height. Evidence of these colossal beings has been discovered in high-altitude caves within the Andes, and species like *Thalassocnus* sp. evolved denser bones, an adaptation that enabled them to dwell in aquatic environments and remain submerged.

This trait has been passed down to modern-day sloths, who are surprisingly good swimmers. They can slow their heart rates to extend their time underwater, and move three times faster in water than they do on land. This ability provides a glimpse into their ancient past and demonstrates how the species thrived over vast evolutionary time scales with minimal changes, reflecting strong ecological niche preferences.

Research into their dental structures and the analysis of fossilized feces have led scientists to conclude that ground sloths were primarily herbivorous, their diets consisting mainly of fibrous plants that varied with the region and ecosystem. They possessed such formidable strength that it's believed they could uproot entire trees to feast on the yummy leaves.

Fossil evidence, including cut marks on their bones made by human tools, supports the notion that early humans hunted these giants or scavenged from their remains. The extinction of these majestic creatures likely resulted from a combination of climate change and increased human activity as humans began to spread across the Americas.

Reflecting on the history of ancient sloths reminded me of one of the most outstanding Peruvian paleontologists I've ever met, Julia. We'd first crossed paths a few years earlier in Manaus, the capital of the Amazon in

Brazil, while participating in an immersive training program on science communication.

Loud chattering and Brazilian samba played in the background as hundreds of people passed by the conference room. The lights were bright and the humidity was intense, with only the faintest hint of a breeze. In the center of the room, I spotted Julia, a short, smart, beautiful, brown-eyed woman standing next to four older male scientists, everyone debating their point of view as fervently as if defending their nation.

— *By the end of the Pleistocene epoch, that's when giant sloths went extinct* — I overheard one of them say.

Immediately drawn in by the concept of giant sloths, I slowly approached, and she nodded to me, inviting me into the conversation. Quietly standing a few steps back, I observed the debate unfold like a fierce tennis match.

— *No, of course not, they must have survived well into the Holocene epoch; there are ceramics!* said an older man, fully convinced of the truth of his statement.

Rock art found in Brazil, Argentina, and other sites in South America depicts figures that closely resemble giant sloths. As interpreted by experts, these images suggest that giant sloths may have gone extinct on the mainland only around 8,000 years ago, rather than 11,000, hinting at a prolonged period of coexistence with humans.

Julia seemed to know this debate could go on for hours, as many theories about ground sloths are still debated by paleontologists today. Standing tall among her peers, her posture exuded authority as she spoke up.

– Well, one thing is for sure – she said. *They occupied all ecological niches out there, from lowland Amazonia to the high Andes—and some were even aquatic.*

– Some people think this furry beauty descends from the mapinguari.

Rufilio's voice brought me back to the field as a family of stunning tiger butterflies (*Tithorea harmonia*) flew by, waving their striped orange and black wings formidably and elegantly.

– But a much less stinky one – he added, waving his hand in front of his nose and laughing as the sloth looked on.

The *mapinguari* was an ancient legend I had heard of as a child, but I hadn't thought about in a very, very long time.

The first time I'd heard about the *mapinguari* was in a fiction book about the Amazon jungle. It portrayed large, one-eyed creatures with brown or red fur that were gentle and benevolent toward humans, serving as guardians to the spiritual world. Yet, they were also depicted as recluses, seeking refuge in the deep, dark caves atop mountains to avoid those who sought to harm them.

A powerful, hairy, and stinky behemoth, the *mapinguari* is said to be generally peaceful but always alert, on the defense, preferring the seclusion of caves, the areas behind waterfalls, or the cover provided by dense vegetation to remain unseen. Some say it has only one eye, some that its mouth is in its belly. Although there are varying descriptions, two characteristics stand consistent: its incredible size (it is said to be capable of felling massive trees with a single push if irritated or hungry), and its potent, repulsive odor, often described as a combination of feces and rotting flesh.

LIVING FOSSILS

Legend has it that the *mapinguari*'s large footprints create a unique sound that is unmistakable: a thumping that echoes through the vastness of the Amazon. This ancestral figure is also said to be very vocal, emitting a terrifying cry and howl that adds to its mystery.

Western science often regards the *mapinguari* as a folkloric explanation for unexplained sightings or unidentified sounds, another intriguing yet unsupported tale to be added to the vast repertoire of Amazonian mythology. It was the "Bigfoot" of South America.

– *What do you think, Rufilio? Do you think they are real?* I asked, curious about his take as we regrouped with the family.

We continued exploring their sanctuary, now heading to the other corner of the land, where a vast pond reflected the sunlight and the nearby elegant orange-colored achira (*Canna* sp.) flowers, which bloomed atop tall stalks with green banana-like leaves.

– *Yes, of course they are real, although I didn't think any were left, not since the time of our great ancestors, to be honest* – Rufilio said.

He paused to pick up some fruit from the ground: perfectly ripe poma rosa fruits. He wiped them with his shirt and offered them around as he continued, his face serious.

– *That was, until a friend of mine heard of someone living on the border between Perú and Brazil who claims to have killed one. He said he shot the* mapinguari *many times in the body, without any effect, almost as if its skin was too thick . . . until he shot it in the head, and then it fell dead.*

Surprisingly, face-to-face accounts of encountering giant *mapinguari*-like beings in the depths of the Amazon aren't rare, and there are a few hunters who claim to have slain one. Irrefutable hard evidence is needed to validate these stories, but the possibility, however slight, of their existence could rewrite the history of evolution in the area.

A month after we'd first met, I spoke about this with Julia.

– The Amazonia was home to a mega-fauna, extraordinary giant animals that have long gone extinct. Yet, some of their relatives are still alive today, and a few still hold the title for some of the largest animals on the planet, like the pirarucu, the capybara, the green anaconda, the manatee, the tapir – she said. She added that our smaller modern-day sloths are a beautiful reminder of the life forms that we used to harbor in our world. *Forget* Jurassic Park—*let's go to the Amazon!*

I laughed loudly, my eyes shining with pride at the incredible history of our natural world.

There are corners of this planet that have yet to be fully explored, where luscious vegetation covers the remains of life that walked there in centuries past: animals that once roamed our jungles, casting shadows across landscapes and shaping life as we know it, their descendants somehow managing to survive the extreme changes and extinctions that we learned about in history and biology class. These beings became the cornerstones of the myths and legends that are woven into the fabric of indigenous cultures across time and space. But they have done so much more than color our imaginations; they spark awe about the potential wonders of nature. These creatures—the pirarucu, the sloth, and other ancient animals I have found—harbor traits that have allowed them to survive through drastic environmental changes that most life forms couldn't withstand, potentially revealing secrets for adaptability in the face of changing climates. It is within these ancient creatures that we can find survival mechanisms to aid in the restoration of endangered ecosystems, guiding conservation initiatives. This is why it's critical to explore the living fossils of the Amazon, and to learn from the local initiatives that are in place to preserve them.

However, living fossils aren't the only source of inspiration for adapting to the extremes. My next expedition takes us somewhere between the Andes and the Amazon, to a place where I have found ancient wisdom that embraces growth in a changing environment.

9

LIVING IN THE CLOUDS

Achiote

SCIENTIFIC NAME: *Bixa orellana*

TRADITIONAL NAME: Achiote, annatto, bija, urucum

ORIGIN: Native to tropical regions in Central and South America, including the Amazon Rainforest

TRADITIONAL USES: Named after Francisco de Orellana, the first European to navigate the Amazon River, this shrub or small tree has traditionally been used by Amazonians to make medicines and textiles. It's also used in spiritual ceremonies, and to make war paint and beauty products. Its seeds, known for their bright red pigment, are used as body paint, mosquito repellent, natural make-up, and sunscreen. Its use as a dye for fabric, cosmetics, and food has skyrocketed worldwide in the last decade, as it is one of the few non-toxic and natural dyes approved by the World Health Organization.

And as a jack of all trades, achiote is also a great culinary ingredient, imparting a yellowish red color and a slightly peppery flavor with a hint of nutmeg. You may find it in butter, mayonnaise, sausages, juices, ice cream, and cheese. The achiote tree is also a powerful native medicine. Its seeds, prepared in a variety of ways, are used as antibiotics, expectorants, laxatives, and antipyretics (to reduce fever), and are also anti-inflammatory. When infused, the leaves aid bronchitis, sore throats, and nausea. When turned into an oil or paste, they can be applied topically to treat skin infections and burns.

SCIENTIFIC INFORMATION: The achiote tree reaches 6–10 meters (20–33 feet) in height when mature. It can be identified by its heart-shaped leaves, soft pink flowers, and spiky fruit pods, which contain numerous red seeds. The pigment found in the seeds is rich in carotenoids such as bixin and norbixin; these are antioxidants, and also provide the red coloring. The seeds are also high in tocotrienols, a form of vitamin E, further contributing to their antioxidant benefits. In fact, hundreds of natural molecules have been detected in achiote seeds, suggesting high biological activity. Research has suggested that achiote may play a role in reducing inflammation, preventing microbial growth and DNA damage, and protecting against oxidative stress. Hypoglycemic effects (a drop in sugar) were also observed in animal testing, but further work is required to understand the impact. Finally, achiote has also been demonstrated to be an excellent natural food preservative.

THE SPIRIT OF THE RAINFOREST

The moon cast a soft glow across my pink curtains as I slowly began my morning routine. Feeling slightly out of place in my childhood bedroom, I felt an odd sense that I had outgrown this space, which was once so familiar. I was no longer the child who had asked her mom to braid her hair before school; gone were the Princess Jasmine leggings and Nintendo Game Boy. Now I was donning my green cargo pants and gathering my science gear—ready to embark on an adventure to the Amazon.

It was just after 4am, and the high-pitched chime of my phone alarm reminded me it was time to prepare for the airport. My internal biological clock was still adjusting after my flight from England to my family home in Perú. I had come for a brief visit before working on some educational exploration trips with the National Geographic Society. As is often the case with these trips, my brain and body were each in a different time zone, and I was the epitome of jetlag. Yet, despite the fatigue, I couldn't help but smile at the thought of returning to one of my favorite places on Earth—a place that continually astounds me with the adaptability of its wildlife.

Descending the outdoor staircase with my luggage, I smiled at the hundreds of paper-like pink flowers, *Bougainvillea,* that adorned the way. I had grown up witnessing this display of flora coming to life every spring. I recalled the first time I'd learned that the pink flowers were actually bracts—modified leaves—that surrounded the true flowers, which were small and white, almost unnoticeable. It had astounded me then, an example of the hidden beauty of our planet. It made me question what else I had not seen yet. And now, as I prepared to return to the rainforest, that same sense of wonder filled me once more.

With the roads deserted, my journey to the airport was swift. I met up with my team, and we embarked on a two-hour flight to the Andean mountains—the gateway to our Amazonian expedition.

We were greeted by the sight of brown, rugged mountains: naked, rough, earthen textures with rocky outcrops. The mountain slopes plunged into deep, narrow canyons, mirroring the sharp decline in my lung capacity as we ascended to the challenging altitude of 3,500 meters (11,500 feet) above sea level.

We explored the Plaza de Armas—a main square central to cities in Perú, named after military parade grounds influenced by Spanish architecture, and now a hub of social life. Boasting a grand cathedral, statues of national heroes, government offices, and various shops and restaurants, the plaza offered rich cultural vibrancy. As we waited for one last team member to arrive—due to the perils of a snoozed alarm—we visited one of the most famous viewpoints in the city.

Under the clear blue sky, we passed beneath tall white arches that stood next to a giant statue of Jesus. Beyond them lay a spectacular panoramic view of the area's wild beauty. Everywhere we looked, our eyes were met by expansive mountains, undulating forms telling tales of the movement and forces that had shaped them. The sun cast shadows and highlights, accentuating the contours of the Andes.

The rugged mountains, occasionally patched by green, are home to wildlife that has tenaciously adapted to the rough terrains and thin, crisp air. This unforgiving landscape nurtures species that have evolved remarkable traits to survive, displaying resilience and adaptability. These, my grandmother always said, are the lessons the mountains teach us.

After a few hours, we were on our way to the Amazon. We boarded a 4x4 truck with our fearless driver, Sergio, who was native to the area and well acquainted with the serpentine roads we were about to follow. It took us over an hour to exit the traffic in the city, and less than an hour after that to leave the paved roads behind. We now found ourselves on a rugged track winding through the Andes Mountains, constantly zigzagging, making sharp 90-degree turns, and unable to see more than a few car-lengths ahead. The narrow, two-way trail hugged exposed rock formations on one side, giving way to sheer precipices on the other: a cliff so tall that it resembled the heights of the mountains, and could worsen any motion sickness if you looked too closely.

This experience brought to mind the European chronicles I had read describing the first Western explorations of the Andes and the Amazon—stories of treacherous mountain paths and dense jungles. It often proved too much for the horses that explorers were riding, which led to deadly cliff falls. Natives are experts in these terrains, and their wisdom cannot be overestimated. Native knowledge has built over centuries to find a balance between human activity and nature, often revealing hidden paths and methods that, in some scenarios, may be lifesaving. This is why, when conducting field expeditions, it is my priority to collaborate with and champion local wise people, like Sergio.

About four hours in, the raw, unadorned terrain began to change, giving way to beds of cacti and dry green and yellow bushes that grew denser over time. As the foliage thickened, the mountains morphed dramatically, displaying the natural union between the Andes and the Amazon.

Along the way, we passed small communities with colorful half-finished buildings and free-roaming farm animals. The road was rough and unpredictable, with skilled Sergio navigating around fallen rocks.

Dusk was falling rapidly, creating a hazy atmosphere with soft oranges, pinks, and purples in the sky, framing the silhouettes of towering trees against the twilight. We stopped at a roadside café, where an Amazonian woman served us some of the region's uniquely aromatic coffee, with rich tones of chocolate and soft spices. After enjoying the coffee and purchasing some to take with us, we resumed our journey, the twisting roads giving way to a lush rainforest. The track was rugged and wild, like something from a *Jurassic Park* movie, and it led us to a clearing. Across the river, distant houses glowed like giant stars under the twilight sky, with "Hotel California" adding a serene soundtrack to the chaotic ride.

– *Welcome to the cloud forests* – said Sergio as he veered right, heading toward the riverbank.

Cloud forests, or "the forests of the clouds," as some may call them, are among my favorite areas of the Amazon. These unique ecosystems emerge where the Andes meet the Amazon: a type of tropical high-altitude rainforest with elevations of 850–2,000 meters (2,800–6,600 feet) above sea level, with some extending into even higher altitudes of over 4,000 meters (13,100 feet). The moisture-rich clouds envelop the forest, leading to average temperatures cooler than lowland rainforests and high humidity. To my delight, these conditions also equal fewer mosquitoes.

Soft light came through the dense clouds, resulting in shimmering shades of hot purples, bright pinks, and dark blues. There was a sense of enchantment across the canopy extending into the sky. Despite their rich abundance of life and unique role in local ecosystems, cloud forests have been largely overlooked by researchers. In these forests, trees "harvest" water from the clouds, acting like sponges, capturing up to 60 per cent of their water intake from the atmosphere. As the condensation accumulates on leaves and branches, it drips to the forest floor and enters streams, contributing to the water that flows down the Andes, the Amazon River, and eventually, into the ocean. This makes cloud forests a major source of water for rivers, with great impact on global weather systems. In fact, if cloud forests were to disappear, a series of uncontrollable floods, landslides, and erosions would take place, leading to human and ecological disasters all the way to the lowlands.

To reach our destination, we had to cross the Apurímac River. Infinite darkness spread ahead, with no bridge in sight. All I could see were thousands of tiny flies and insects around the car that had been attracted by the light.

– *They've closed the* chimpa[1] *(crossing)!* exclaimed Sergio in exasperation, his panic taking us by surprise. *During the day, there are floating platforms available to take us to the other side . . . but they are gone!*

– *What's our next option, Sergio?* I asked in a low, calm tone. Internally, I wondered how in heaven we were supposed to cross a pitch-black river in a truck, but I invoked the lesson Sergio himself had taught me on a previous trip: calm, control, and order.

– Doctora, *the nearest bridge from here is over three hours into the deep forest* – he responded, biting his lip.

Our team member Cesar shook his head. He knew the territories nearby could be very dangerous when you were approaching unannounced, particularly in the middle of the night.

As we contemplated our options, a quirky romantic ballad started playing from the car radio.

– *Could we camp out here?* I said softly, knowing very well that meant sleeping in uncomfortable positions in the car, without any camping gear on hand, exposed to the wilderness of the rainforest. The nearest houses we could see were across the river, and the distance was misleading, as they were probably many miles away.

Just as the tension began to mount, a light illuminated the river ahead, revealing an unusual-looking hybrid of a moving platform and a boat. As it made its approach, a young man in shorts walked toward us, asking if we needed to cross. A collective sigh of relief filled the car as Sergio shook hands with the man then drove the truck onto the platform.

Although I understood the logic of what was happening, the minimal lighting made it seem like we were driving straight into the river. My heart squeezed in my chest as the song on the radio increased in intensity. Crossing a tributary of the Amazon River in the dead of night, inside a truck that was perched precariously on a moving platform, was one of the most unexpected rides of my life. In the distance, animal sounds resonated through the forest, amplified by the echo of the valleys. The dramatism of the situation contrasted with the sluggishness of the movement. The motor roared. The water split open. And we sat in the stationary truck until the platform slowly kissed the other side of the riverbank with the speed of a sloth reaching out between trees.

After crossing the *chimpa*, we paid the man for his service and continued on our way. It was too late to enter the community we were there to visit, so we found the closest hostel and ordered giant plates of *chifa*, a signature Peruvian-Chinese dish found throughout the country, which my British husband describes as our own version of fried rice, but less greasy. It's an ageless delicacy that never fails to satisfy. As Latino culture says, *"Barriga llena, corazón contento"*—a full belly equals a happy heart. Finally, we rested for the night, contented and filled with warmth.

The next morning, we made our way toward the Ashaninka community I had visited previously; my heart was bursting with excitement at the thought of seeing Pascual and Micaela again. This community is particularly special to me given their spiritual interpretation of creation and the afterlife, and the fact that my great-great-grandparents also originated from the high-altitude Amazon Rainforest. The *Avireri* mountains welcomed our return, filling the misty air with green reflections of their infinite flora. We heard the echoes of the synchronous drumming, flutes, and singing before we had sight of the community.

As we slowed the car, dozens of men, women, and children of all ages emerged in the distance, marching together toward us. Each was adorned in a unique *cushma* (traditional clothing), in shades of orange or off-white with thick brown lines. The fibers of the *cushma*, when first obtained, boast a natural off-white tone, which some prefer to dye using tree bark, traditionally known as *pochotakori*, and other natural resources, resulting in the colorful and eye-catching orange and brown hues.

The Ashaninka men and children carried quivers filled with hunting arrows, and wore necklaces of native seeds and dried fruits. At the center, Pascual sang joyfully while playing a drum crafted from animal skin. He walked alongside Marina, his mother, who wore seed arm pieces that chimed like rain. At the *maloca* (community center), I met Estela, the local primary teacher, a woman with strong eyes and a kind spirit. With an achiote fruit opened in one hand and a small natural brush in the other, she started creating unique paintings on our faces.

– *A cat* – she whispered to me. *You have the spirit of the cat.* She was referring to the jaguar. Estela looked deeply into my eyes, as if scanning me from inside out. Her face muscles relaxed as she smiled. This ancestral welcoming ritual of singing, dancing, and painting invoked the *Avireri* to watch over us during our stay.

Immediately afterwards, a bowl filled with *masato* (a traditional drink, see page 27) was passed around for each guest to drink, symbolizing familiarity and trust. Every Ashaninka man and woman was in high spirits, delighted to see us arrive. They explained that fewer people now visited, as their town was located at one of the last habitable spots in the intersection of the Andes and Amazon, the most remote route on land. In fact, it was one of the very few cloud forests remaining in all of Perú and the Amazon Rainforest.

Catching up with the family felt like reconnecting with old friends, sharing warm laughs and personal stories. New faces had joined the group—Ashaninka young professionals who had mastered Spanish and completed their formal education, and were now serving as cultural representatives and park rangers within the local Ministry. These roles were crucial in managing the National Reserve that bordered their land, a title that, while providing a degree of respect and protection, paradoxically also attracted relentless threats from those intent on exploiting the rainforest for wood, mining, and other destructive activities that might put at risk the uniquely adapted wildlife of the rainforest.

I shared with Pascual and everyone at the table our eagerness to blend scientific exploration with indigenous knowledge in order to bolster conservation efforts. My focus on the unique adaptations of flora and fauna at 1,500 meters (5,000 feet) above sea level—including their uses in traditional medicine and their spiritual significance—highlighted the potential for merging the scientific and indigenous worlds.

The Ashaninka community's adaptability in response to modern threats reflects their embrace of science. They recognize the potential of scientific advocacy, witnessed in other parts of the rainforest, to secure policies that promote long-term nature preservation.

Cloud forests, constituting only 1 per cent of the world's woodlands, are rich in endemic wildlife found nowhere else on Earth. Each elevation creates its own microclimate, harboring unique plants and animal life. It's an evolutionary gift to humanity: wildlife knows no borders and adapts in unpredictable and beautiful ways. Just think of hummingbirds, which require vast amounts of oxygen for their rapid wing battering and intense cardiovascular activity, yet have evolved to thrive in these oxygen-poor environments. Or stingless bees, which flourish in the lowland floodplains of the rainforest, yet have also learned to adapt to the high altitude of the cloud forests.

The physiological and behavioral changes that the local wildlife undergoes—including surviving at reduced oxygen levels, lower temperatures, and in different types of vegetation—hint at a unique interplay between genetics, microbiomes, chemistry, and environment. This journey promised new knowledge that could expand our understanding of biodiversity and perhaps open the door to innovative ways of protecting it.

Together with our Ashaninka guides, we grabbed our science gear and rubber boots and headed into the deep jungle. We were on the lookout for wildlife that had evolved their metabolic systems to survive and thrive in a drastically different ecosystem, speaking to their rare and remarkable resilience and versatility.

After 30 minutes of hiking, we took a left turn in the forest. Leading the way, the Ashaninka began to ascend a challenging, muddy slope, inviting us to follow. Within moments, the rapid ascent quickened my heartbeat, demanding deep breaths to steady the rush of high-altitude adjustment. My hands felt extremely cold as I grasped the nearby lianas cautiously, avoiding hidden dangers. I became slightly dizzy, prompting a brief pause and a thoughtful offering of lime-flavored candy from one of our Ashaninka friends.

As we navigated through the dense underbrush, the terrain altered constantly—roots sometimes served as makeshift stairs, while mud clutched at our boots, slowing our pace. During the ascent, Pascual's voice broke through.

– Doctora, *I don't know if I've already told you this, but we have run into bears here before* – he said, as he agilely hiked up one of the steepest sections of the path with minimal effort.

– *Bears as in . . . Andean bears?* I asked, intrigued, trawling through my memory for the only bear species I knew of that lived in Perú.

This spectacled bear, with white markings around its eyes, is the only native species of bear in South America, and was the inspiration for the origin story of Paddington Bear, who came from "darkest Perú." As the name suggests, Andean bears primarily inhabit the Andean mountains. However, right there and then, I learned that these black- or brown-coated bears have also adapted to thrive in the Amazon cloud forests— the very high-altitude areas in which we were now hiking.

– *Yes, black bears live in packs around here. I saw a mother with her two cubs just recently* – Pascual continued casually, his voice indicating complete calmness and normality.

– *Are there any precautions we must take if we run into some?* I asked. I thought I could predict what he might say, but the left side of my brain whispered a contrasting thought: *Run.*

– *Bears are wise and respectful; if we show them the same respect back, of course* – Pascual said. *When we run into them, we simply acknowledge each other, and then off we go on our own path, and off they go on theirs. Even the mother bear with her cubs, she knows we never hurt them or hunt them.*

As Pascual spoke, his eyes signaled a deep sense of admiration for these mammals.

– Bears have a strong spirit; they embody wisdom, courage, and strength. They are powerful protectors of the Avireri, *constantly embracing change and rebirth* – he added.

I like to think of this moment as one of those times where life aligns to reveal patterns that may not have otherwise become apparent—little bites of wisdom that beautify our view of the world. Pascual's words resonated deeply with both my indigenous ancestry and my scientific mind. Having recently learned more about the Shoshone and Bannock indigenous people who used to live in Yellowstone, as well as the Yup'ik and Dena'ina in Alaska, I connected the dots to realize that all these indigenous communities, like the Ashaninka people in the high-altitude forests of the Amazon, view bears through a very similar ancestral lens. Symbolizing strength and intelligence, these animals are revered for their power and connection to the land, inspiring respect.

They are extremely different ecosystems, yet a common thread connected the Amazon, Yellowstone, and Alaska—the indigenous significance of bears. This ancient wisdom has shaped the adaptability of local people, who have learned to live harmoniously with wild creatures. This stands in stark contrast to modern measures like the imperative use of bear spray in national parks for hiking, fishing, wildlife viewing, or photography. While necessary for safety, such measures underscore how our exploitation of their lands has exacerbated many aggressive interactions.

Pascual's words encouraged me to reflect on the possibility of peacefully coexisting with nature, of seeking a deeper understanding of wildlife to minimize conflict and live more harmoniously. And the first step to this is to acknowledge and elevate indigenous wisdom using scientific platforms that guide policies and action.

This idea was vividly illustrated in the days that followed as we immersed ourselves in the Ashaninka territory. Our team worked closely with emerging local field scientists, employing modern technology to monitor biodiversity and gather essential data. This collaboration, deeply enriched by the Ashaninka's profound connection to their environment, led to the publication of a significant ethnoecological research paper detailing Ashaninka knowledge regarding local fauna, marking a historic moment. It was one of the first instances in Perú when indigenous community members were acknowledged as co-authors, recognizing their invaluable intellectual and practical contributions. In recognition of my efforts, I was honored with the title of International Ambassador of the Ashaninka. This endeavor not only bridged the gap between traditional indigenous knowledge and contemporary scientific research, but also set a precedent for integrating indigenous insights into scientific studies at all levels. Ultimately, indigenous worldviews are largely inspired by observing, understanding, and adapting to the complex patterns of nature, showing a natural inclination toward the scientific method that has long gone ignored.

The cloud forests of the Amazon are also home to other extraordinary wildlife: mountain tapirs, woolly monkeys, and other key species that act as biosensors, monitoring the health of the environment. Interestingly, cloud forests have become the last refuge for some species, preventing them from entirely disappearing from the Earth's surface. With temperatures increasing globally, various life forms, including trees and beetles, have migrated to higher elevations to maintain their optimal temperature. It has been observed that the tree line has been moving up about 2.5 meters (8 feet) a year, while some beetles have already moved up 40 meters (130 feet) a year—an ongoing phenomenon that shows wildlife adapting to avoid extinction.

As we drew close to the town, some Ashaninka community members approached us, holding tiny spotted dark green frogs in their hands. Pascual had instructed them to collect a few of the native frogs in the area to show us upon our return, sharing with us one more ancestral practice they maintain to this day.

I was immediately intrigued. From all the frog species found worldwide, only a small subset adapts to high elevations due to the challenging conditions. Those that do so must change their breeding cycles to adapt to the shorter warm seasons, reduce their energy use, and improve their oxygen uptake, potentially changing their hemoglobin structure. Their presence in these lands speaks to their remarkable ability to acclimate.

– *These are poisonous frogs*, doctora – Pascual said, carefully clutching one tiny specimen in between his hands. I could see the frog's body expanding and contracting with each gentle breath. Noticing my eyes grow wide, he laughed and added – *They only release their poison when stressed; they are calm around us.*

Pascual gazed at the frog with the same eager fondness I reserved for my tiny puppy back in England. Our photographer captured close-up shots of the frogs, as Pascual wanted to have them readily available to share. As we watched the frogs calmly pose in front of the camera, without any flashes to avoid stressing them out, Pascual continued to explain the Ashaninka tradition. He shared that they had learned from their great-grandparents how to remove the poison from the frogs' bodies, making them safe to eat during seasons when they breed extensively or in periods when food is scarce for the community. It was another example of the Ashaninka's resilience—living in cycle with the Earth's rhythms. Their adaptability comes from centuries of close observation of and respect for nature, through which they have learned how to meet their needs without jeopardizing the wildlife around them.

My mind buzzed with questions. From my scientific background, I knew that poisonous frogs have long been a source of potential new medicines. For instance, the potent analgesic drug epibatidine was discovered in the skin of the *Epipedobates tricolor*, a poisonous frog from the rainforests of Ecuador. This discovery was inspired by the local use of frog secretions for hunting, where they were applied to darts or arrows to quickly immobilize prey. Epibatidine represented a breakthrough as a potentially more effective analgesic than morphine, but without the addictive side effects, paving the way for new methods of pain-management.

At the same time, my indigenous knowledge reminded me of practices in remote Amazonian communities where individuals burn a small area of skin and then rub a native frog called a *"kambo"* (*Phyllomedusa bicolor*) on the raw skin. The frog toxins are quickly absorbed transdermally into the bloodstream, leading to vomiting and physical pain after minutes of serenity. This deeply spiritual ancestral ritual is aimed at purifying the body and the mind. Practitioners have claimed that it can cure depression, drug dependency, high blood pressure, and other issues. However, there is no solid research or legislation surrounding *kambo* treatments, and severe cases of intoxication have been previously reported.

Interestingly, with over 300 recognized species of poisonous frog in the Amazon Rainforest with unique markings, electric colors, and toxic skin, a study revealed that 80 per cent of their ancestry originated from the high Andes Mountains, which served as an oven of evolutionary biodiversity. Unfortunately, one in three types of amphibians in the region is now threatened with extinction.

But in between all the stories, tales, and research, I had never come across the knowledge that Amazonians have learned how to extract the toxins from poisonous frogs to make them safe for human consumption. How does that work? Can you really extract all the toxins? Has the human microbiome evolved to prevent intoxication even if trace amounts of poison remain? Some birds, reptiles, and insects, like elegant praying mantises, can consume venomous creatures without harm, exploiting a niche into which other predators cannot venture. This speaks to their unique adaptability and potential microbial interplay. Similarly, the human microbiome must have adapted in unique ways in order to digest unconventional food sources, such as toxic frogs.

The next day, Pascual announced that the community's children had invited us to visit their primary school, which was located a kilometer away from the huts, toward the green mountains. Estela, the esteemed Ashaninka educator who had painted my face during our welcome ritual, warmly invited us to join one of their morning sessions. Stepping inside, we found a dozen Ashaninka students, aged between six and twelve, seated at small wooden desks, clad in their *cushmas*, adorned with achiote paint, and with bows and arrows slung over their backs. All were attentively focused on the chalkboard at the front of the room.

As I saw the students' sweet faces smiling, I couldn't help but think of my grandmother. Her family had migrated from the Amazon to the Andes one generation prior, but she'd always dreamed of attending school. In fact, for a few months in her early 20s, she would sneak out at night after tucking my mom and aunt into bed. She was careful not to wake my grandfather, who was exhausted from a long day shaping wood into doors or coffins, or whatever else was on demand. She met with other young women in the town by the dim light of a small candle. They had collected discarded magazines and book pages from the trash during the week. Together, they were trying to teach each other how to read and write, claiming their strength and independence, and broadening their knowledge. Their secret, improvised school continued for a few weeks until the flickering flames of their candle alerted one of the husbands. Fearing that the women's interest in reading and writing signaled a desire to leave their families in search of younger men, the men put an end to the lessons.

Indigenous education is crucial, especially in remote corners of the planet where insights into the wider world are scarce. My grandmother's brief experience of self-learning through magazines fueled her determination to ensure future generations had unrestricted access to education, empowering us to achieve our dreams.

Now, eager to showcase their abilities and culture, the class led us outside to share their version of Physical Education. The boys formed a line and marched toward the target, a round wooden bull's-eye set against a sturdy tree fiber post. Taking turns, they each took aim with bows in hand, pulling back the string with intense focus. A brief hush fell as they reached full draw, and then with each release, an arrow whizzed through the air—some thudding into the target, others burrowing into the forest hill, and a few striking true at the center.

Then, in an unexpected twist, the younger girls stepped forward, grabbing bows and arrows with confidence and precision. In most rural communities of the Amazon and Andes, women are not encouraged to learn how to hunt. Those are activities reserved for men, just as reading and writing was in my grandmother's time. The youngest, a six-year-old girl with a long *cushma* adorned with drawings of parrots and plants, gave a big smile before momentarily turning serious, her eyes focusing forward. In the length of a breath, she pulled back and threw the arrow with the highest precision of all, hitting the bull's-eye. Breaking the silence that had built up in anticipation of what might happen, we all screamed in excitement, with the girl doing a little jump before letting her best friend go next.

My heart filled with pride.

Pascual revealed that, just like the wildlife around them, their Ashaninka community also embraced change. Here, each child was encouraged to pursue their passion, unbound by traditional social or gender expectations. Girls interested in learning hunting skills were welcomed, just as boys who showed interest in sewing with fibers and seeds were encouraged. In this stage of evolution, roles and expectations were not imposed, but rather chosen, honoring the individual's interests as the community evolved and strengthened. It is a lesson that could inspire humanity to embrace growth from within.

During our stay in the high-altitude rainforest, we also spent extended time with Pascual's family, who kindly showed us the various ways in which they prepare medicinal plants inside their traditional kitchen. At one point, we witnessed his father emerging from the deep forest with a large white bag slung over his back. It was filled with hundreds of medicinal leaves and flora. With the *Avireri* on one side and the expansive rainforest behind, the scene evoked memories of the old biochemistry books in which I had first encountered photos of scientists traversing lands and collecting thousands of unique medicinal plants to study in laboratories. This research often led to the discovery of or inspiration for pharmaceuticals we still use today. In the case of Pascual's family, the plants represented their natural pharmacy.

– *He went to visit his father, the* sheripiári – said Micaela, who was standing next to me.

I remembered Pascual's stories of his grandfather's spirit, now residing within a tree that provided infinite sources of medicine for their family.

Amid the clanking of pots filled with healing herbs, the fire crackled, heating special rocks from the *Avireri* as the logs burned. This released a plethora of aromas that perfectly complemented the scent of the various medicinal roots, leaves, flowers, and plants that Pascual's father had gathered from the rainforest. As we savored these aromas, we were immersed in the indigenous knowledge of the high-altitude rainforest, exchanging stories, dreams, and jokes, reveling in each other's company and bridging our worlds.

Through the Ashaninka lens, I saw that nature in the cloud forests embodies adaptation with mastery and precision—neither too slowly nor too quickly, but at a perfect self-determined pace. Each species manipulates its own genetics and chemistry to embrace change within its own circumstances, evolving to belong.

The high-altitude rainforest taught me that nature has a way to preserve its heritage by connecting us to the past and offering a glimpse of what the future could look like. Inspired to explore how nature and people can coexist mindfully of one another, I delved into the study of ancestral civilizations, including those long vanished.

10

ANCESTRAL CIVILIZATIONS

Cinchona tree

SCIENTIFIC NAME: *Cinchona* spp. (encompassing approximately 40 species)

TRADITIONAL NAME: Quina

ORIGIN: The foothills of the Andes down South America, including high-altitude cloud forests where the Andes and Amazon Basin converge

TRADITIONAL USES: Esteemed as the national tree of Perú and Ecuador, the cinchona tree has served as a pivotal medicinal resource in the Andes and Amazon for centuries. Long before European colonization, Quechua communities harnessed the bark of this tree to alleviate inflammation, fever, pain, and infections, effectively treating malaria. The traditional technique of grinding the bark and mixing it with sweetened water provided relief from shivering, acting as a muscle relaxant. Curiously, quinine, an essential bittering agent in tonic water that gives the drink gin and tonic its distinctive taste, is commercially extracted from this tree's bark. Moreover, the bark is also celebrated for aiding digestion and strengthening gastrointestinal health. Historically, quinine extracted from the bark was also used as a natural dye, producing shades of yellow to reddish-brown.

SCIENTIFIC INFORMATION: Identified by its glossy green leaves and flowers, which vary in color from pink and white to yellow, and perch atop a slender trunk, the cinchona tree can reach heights of 15–20 meters (50–65 feet). Its bark is laden with alkaloids, most notably quinine, which interrupts the life cycle of malaria-causing parasites within human bloodstreams, marking the dawn of a new era in the treatment of malaria. Quinine's analgesic, antipyretic, and anti-inflammatory properties have secured its place on the World Health Organization's List of Essential Medicines. Historical overharvesting of the tree highlights the critical need for sustainable management practices to safeguard these invaluable species and their surrounding ecosystem.

I couldn't shake the feeling that something—or *someone*—was following us.

There was a shadow that mimicked our movements, an eerie echo in the distance. It glided through the dense foliage gracefully, methodically matching our speed and direction, as if tethered to our every step.

We had been navigating the Napo River for some time, surrounded by vast expanses of green and trees of every size. There was not another human or animal in sight—until that unsettling glimpse of the shadow.

As I squinted to get a better look, trying to decipher its form, the shadow abruptly vanished into the dark forest. The sunlight, reflecting off the water's surface, pierced through the evergreen canopy and blinded my view, trumping my efforts to track its path.

My eyes stayed fixed on the spot where the shadow had disappeared, but the forest revealed nothing more. After a few minutes of intense gazing, specks began to dance across my vision. I decided to relax my eyes, dismissing the shadow as perhaps a monkey or a wild cat playfully leaping among the tree branches. But the unsettling feeling lingered.

Our boat pressed on, deftly avoiding the lianas that intertwined beneath the water's surface as we made our way along a languid tributary. Around us, dozens of light-green water lettuce plants (*Pistia stratiotes*) floated, adorning our route. The air felt crisp, fresh, and invigorating as I closed my eyes to etch the beauty of this journey into my memory. Barely five minutes had passed when a series of sounds suddenly erupted.

Dum.

Dum.

Dum-dum-dum.

Dum-dum.

Dum...

Deep, hollow, and far-reaching drumming that startled us with its intensity.

It pulsed in a repetitive pattern that resonated with almost primal familiarity. The drumming felt unexpectedly soothing, as though this wasn't the first time I'd heard it. Except that it was, and we were right in the heart of the Amazon. My brain fired up, instinctively discerning that this drumming was either a welcoming sign or a threatening one. It became clear that the shadow hadn't been a figment of my imagination.

Less than a minute later, the drumming subsided, amplifying the ambient sounds that had accompanied us for hours: the motor's hum, birds in flight, distant howling, water leaping and splashing against our boat. The deep drumbeats left an echoing void in the atmosphere.

With no clear space along the riverbank to allow us to pause and assess our situation, we had no option but to continue following the water's course. It was unclear whether the drumming was for us, and whether it had been a greeting or a warning.

As we took a left turn, the brown, sediment-rich waters grew shallow. Far ahead, a lengthy wooden staircase stretched down from the towering riverbank. And at the top stood a Bora man, waving at us with a big, welcoming smile.

Deep relief washed over us, and we released a collective sigh. We were OK.

My lungs filled with air once more, and as the last ounces of anxiety dissolved, we laughed and waved back, feeling reassured and relaxed.

When we'd heard the drumming, our first thought had been that we had inadvertently trespassed on someone else's private land—a misunderstanding that could have easily arisen given the region's history of colonization and exploitation. This is why some Amazonian communities can be wary of unexpected visitors: such history prompts a defensive stance, where communities are ready to protect their own. To this day, my grandmother keeps a close watch on any unidentified car that may park across from our home in the city for longer than an hour.

This experience could have quickly turned into a rather perilous situation, where we stood no chance against an attack of poisoned arrows or rocks. We were exposed in the middle of the water, in a small, open boat. I realized we didn't even have any substantial bargaining tools, which such a situation might have demanded. I recalled a conversation with a scientist of Amazonian descent who mentioned he always brought Inca Kola with him—a yellow carbonated soda known as "the Peruvian Soda"—when establishing relationships with new communities. Revered as a national favorite, this soda could be perceived as a token of goodwill by Amazonians. Yet, with no Inca Kola on hand, our possessions were limited to reusable water bottles and a few small banknotes tucked into our pockets.

We rowed the last few meters to keep vegetation or branches from jamming the boat's motor, with the sun now searing our skin as the clouds that had accompanied us all morning quickly dissipated.

– *O tsájucóó!* – the man, Eli, greeted us with the energy of a long-time friend, though it was our first proper meeting. Our collaborators had kindly put in a good word for us, securing an invitation to visit their community. We were here with a local biologist and an international artist to explore their way of life and the surrounding biodiversity. They were already familiar with our work and intentions, and were eager to guide us through their world to strengthen their conservation and communication efforts, ready to share their ancestral Bora practices with the world.

The Bora are an indigenous group primarily located in the Peruvian Amazon, with additional communities in Colombia and Brazil, and they are a living testament to resilience and endurance. During the 19th century, driven by the Industrial Revolution and the increasing demand for rubber for the automotive industry, many people came to the rainforest to set up networks of rubber-extraction operations. Essentially, the Amazon served as ground zero for the rubber boom. The operations were ruthless, and caused widespread destruction that forced groups like the Bora out of their lands. Rubber tree species were exploited indiscriminately for latex extraction, and indigenous peoples were often enslaved under harsh conditions, with violence used as coercion. The rubber boom's decline in the early 20th century gradually allowed communities to regain control of their lands and revive their ancestral traditions.

The Bora are known for their rich ceremonial life, which includes body painting, traditional wear, dances, and rituals. This is how they honor their spiritual and ancestral worlds, keeping their culture alive and honoring the biodiversity around them that is part of their daily lives and also of great spiritual significance.

Barefoot, Eli donned a beautiful off-white skirt adorned with geometrical patterns of long black lines meant to embody the shape of an anaconda. For the Bora, the anaconda is a powerful spirit that represents transformation and healing, and is considered the creator of local waterways. Eli's garments had been crafted with fibers of *llanchama*, a cloth derived from the bark of the *Ficus* tree. The bark is sustainably harvested, ensuring the tree remains unharmed, and is beaten until it softens and becomes pliable, transforming its texture and turning it into a durable material without the need for weaving—a prime example of sustainable fashion.

The fabric is then dyed with *huito* (*Genipa americana*),[1] one of my all-time favorite fruits and natural resources from the Amazon. When still green and unripe, the juice from the pulp serves as a potent natural dye, used both for body painting and coloring fabrics.

I was about ten years old when I first encountered *huito*. My family and I were guests of a remarkable matriarchal community in the Amazon. They welcomed us, eager to share their practices and traditions. I was struck by the women's auras of strength and grace; they stood tall, their long, straight hair a deep black, a color maintained even by the community's elder leader. As I touched my own black hair, I hoped it would retain its color so well into my own later years: a sign of femininity that, in some Amazonian cultures, also symbolizes strength and social unity.

One afternoon when we were with that community, I witnessed a captivating tradition: elder women were decorating the hands and arms of the younger adults with temporary tattoos made from *huito*. The women intricately painted the girls in a chain of beauty and heritage among Amazonian women. It reminded me of my grandmother, mom, aunt, and cousins when they would braid each other's hair, while exchanging stories and laughter.

I grabbed one of the hard, green oval-shaped fruits, looking to access the mysterious black ink. One of the community leaders, with infinite patience and a long smile, sat next to me. Slowly, she opened the fruit with a small knife, revealing what looked like an off-white pulp and seeds. She laughed, seeing the confusion in my eyes as I tried to understand the difference in color. If the ink was black, how come the inside of the fruit was white?

The woman added a few drops of water to the pulp, vigorously stirring it with a small stick. As the pulp became more liquid, it magically changed color, going from off-white to black. I later learned that as the juice is exposed to air, it oxidizes, changing color. Chemistry at its purest.

She drew a fascinating combination of lines and geometrical shapes on my hand that reflected the spirit of birds, as a tiny hummingbird curiously flew around us at the time, an omen of good luck. This temporary tattoo lasted for a few weeks as I carefully protected it, wanting to immortalize the experience. (I learned that Amazonian women use this ink to naturally dye their hair black, and it lasts for months at a time.)

Eli helped us tie our boat to the wooden staircase, ensuring it wouldn't bob away. As we secured our ride, the rainforest surrounded us with its dense vegetation and immersive sounds. We were now deep in the Amazon. Though Bora is his primary language, Eli speaks fluent Spanish and acts as the correspondent between his community and neighboring cities. I noticed his cheeks were adorned with two horizontal lines of *huito* paint, a traditional mark of identity and status within his clan, as well as a talisman against evil spirits, safeguarding him during hunts and daily life. He also wore an elaborate and majestic headpiece made with feathers from the brightly colored *guacamayo* (macaw) and other bird species, the colors and patterns carefully selected to display a symbolic role of leadership. At the back of the headpiece, two long, orange feathers soared. These feathers are often collected as they are naturally shed by the birds, respecting the rainforest's biodiversity and natural cycles.

As we followed the trail toward the center of the town, several species of birds and butterflies appeared from within the dense forest, adding a plethora of colors that momentarily decorated our surroundings before they quickly flew away. A charming bright yellow cloudless sulfur butterfly (*Phoebis sennae*) approached us very closely with its wings moving slowly, before rushing away as a large iridescent blue morpho[2] appeared behind us, the pair of them creating a trail of yellows and blues that dissipated into the evergreen of the forest. The morpho butterfly's iridescent blue wings reflected the sunlight with their microscopic scales, casting a magical aura over our trail as if we had stepped right into *Avatar*.

– *Our families have prepared a welcome ceremony for you all* – Eli revealed eagerly as we continued trekking with growing excitement.

After a 30-minute hike through the forest, the trail grew wider, and we moved into a large, circular clearing surrounded by towering trees. At its center, a pathway laid with long wooden planks led to the *maloca*. Its impressive size and distinctive conical roof were carefully designed to prevent leaks during the torrential rains of the Amazon. This structure, with its wooden beam framework tied together with natural fibers, served as a gathering point for all social activities, including ceremonies.

Eli's brother waited by the entrance to the *maloca*. He greeted us happily, gesturing for us to enter and enjoy the shade provided by the thatched roof. The sun blazed overhead in the clear sky. He spoke a few words in Spanish, each followed by a hearty laugh. My eyes shined and I smiled back at him, knowing exactly how he felt when speaking this non-native language. It reminded me of the first time I moved to the USA for my undergraduate studies; I'd felt as though I had forgotten all the English I had learned, and found it hilarious that foreign words were coming out of my mouth and eliciting a response from those around me. I recalled thinking that if someone talked to me, I would just smile and say "yes," hoping they hadn't asked me anything important.

Upon stepping into the *maloca*, we were greeted by its expansive interior. There were no internal walls, creating an open and spacious environment. Strategic openings in the exterior walls and roof invited a soft glow that illuminated the compacted and smooth earth floor beneath us. Along the sides, long wooden benches and tables offered spaces for rest, while the opposite wall was adorned with an array of artistic creations and decorations, showcasing the Bora's exceptional artistry and craftmanship.

To my right, I noticed two thick, cylindrical, hollowed-out dark brown tree logs—the Bora's version of drums. Suspended by ropes from the ceiling and positioned diagonally, they were supported by an ingenious wooden framework. The logs each had a smooth surface and two rectangular holes artistically carved at the top and the bottom. Each side of the holes featured a half-circle shape, interrupting the straight edges of the rectangle with concave curves. A slender, deep slit linked the two openings.

– *I am going to announce your arrival to the rest of the families, so they all know it's time to gather and we can begin the ceremony* – Eli explained, as he picked up two maraca-shaped wooden instruments, also known as beaters or mallets, that lay atop the logs.

Intrigued, I observed as his expression shifted to one of concentration. He began rhythmically beating the logs, each arm and beater dedicated to one log.

Dum-dum-dum.

Dum-dum.

Dum.

Dum-dum-dum.

Dum-dum.

Dum.

The powerful, resonant, and low-frequency drumming vibrated in our chests, echoing through our bodies and into the forest's depths. This was the same bass tone that had alerted us earlier, yet now it followed a very different rhythmic pattern.

The *manguaré*, known as the heartbeat of the Amazon, is a traditional instrument that is used as a unique form of communication both within and between communities, with its sound reaching over 20 kilometers (12½ miles). No wonder we could hear it so vividly from the middle of the river, feeling the drumming like a pulse. For comparison, human speech has a maximum range of about 200 meters (650 feet). Thus, these drums enable the Bora to transmit messages across great distances. The messages can vary widely, from asking someone to bring you something, to organizing meetings, issuing alerts, announcing results of a competition, or signaling the arrival of visitors.

Each drumbeat mirrors a syllable of spoken Bora, with announcements typically comprising approximately 60 drumbeats to convey about 15 words. Scientific research has demonstrated that the drumming closely imitates the tone and rhythm of the spoken Bora language, whose rhythm and length of pauses between beats are key to distinguishing words. Essentially, the Bora have masterfully encoded their entire language into drum patterns so their words can travel effectively through the depths of the Amazon. Currently, only about 1,000 people in northern Perú still speak Bora, classifying it as an endangered language.[3]

– *The* manguaré *was here before my great-great-great-grandparents, and I hope it's here for my great-great-great-grandchildren* – Eli said.

Complex languages are a key component of advanced civilizations. They serve as the essential foundation for the development of technology, trade, and governance. For example, the Inca civilization did not have a known written script, but developed the *khipu* system, a recording apparatus comprised of strings with knots in varied colors and patterns. This method represented a sophisticated approach to counting, enabling the Incas to conduct detailed censuses of their population and manage crop production throughout the Andes. While still a topic of scholarly debate, some researchers speculate that *khipus* may have also encoded linguistic information and stories, which remain to be decoded and understood.

In fact, chronicles from the explorer Francisco de Orellana, who ventured into the Amazon in the 1540s, depict highly organized and advanced societies. These accounts describe hundreds of settlements housing thousands, perhaps even millions, of individuals. Such chronicles are among the first known written records of the Amazon Rainforest.

These historical accounts also frequently mention hearing well-attuned drumming from indigenous communities, sounds that carried over long distances. This drumming was used to signal war, summon gatherings, and communicate vital information, suggesting that such practices were millennia old, predating the arrival of Europeans and woven into the Amazon's social fabric as crucial survival mechanisms. It's plausible to suggest that the Bora's *manguaré* drumming language could have roots in, or connections to, these ancient practices.

Similar practices are found throughout the Amazon Rainforest. These range from using flutes whose melodies mirror the structure of language, to the art of whistling messages to communicate without disturbing prey, and even incorporating symbols from *ícaros* (healing songs) and messages into textiles. There is a plethora of profound cultural knowledge that underpins the worldview of indigenous groups and their unique understanding and interconnection with the planet.

These ancient practices are testament to the fact that civilizations in the Amazon had the wisdom to develop languages in balance with the natural world, and which serve as an extension of their communication, allowing them to live in harmony with nature for generations.

– *Sit, sit, please, it's about to begin.*

Eli ushered us swiftly to the long wooden benches on the right as community members of all ages came pouring in, each sporting a unique pattern of *huito* paintings on their clothing, and wearing a variety of colorful feather headpieces. Some clutched canes as tall as themselves, topped with small pouches of dried native seeds that rattled with the taps of each cane against the ground.

The air was filled with the sound of singing in the Bora language. As everyone danced in a circle, the rhythmic rattling of the seeds accompanied their movements. At the circle's heart, a woman with a child in her arms swayed to the music. The elder men, including Eli, extended their left arms to rest on the right shoulder of the man in front, moving in harmony around the circle's edge.

Singing proudly and with a bright smile, Eli signaled for us to come join their dancing. An elder woman, her smile so wide I could count her teeth, took my hand, helping me integrate into the group's synchronized stepping pattern. I instinctively smiled back as I thought that here, within the depths of the rainforest, lay the key to a better future for our planet. I was in awe of the Bora's resilience, the way they had bounced back from severe persecution without jeopardizing their spirit of collaboration for the betterment of their society. Their spirit of forgiveness, their determination to rise above one of the most brutal invasions and exploitation campaigns in the region, radiated around them, instilling a sense of compassion and hope. The genuine joy within the community was a testament to their generosity and openness to sharing their rich cultural wisdom. It is within the Bora's worldview that we can find inspiration for living in deep connection with the environment; by embracing forgiveness, we can find peace with one another.

After an incredible ceremony, and time shared together learning more about their artistry and practices, the Bora escorted us back to our boat. The sun was setting, and we still had to navigate a few hours upriver to return to camp. A palpable energy of warmth and camaraderie filled the air as everyone hugged each other. We had formed a close bond in a short amount of time.

– *Our drumming language is older than our trees,* doctora – Eli said wisely as he returned to our conversation with profound insight. His children laughed and ran after a family of Amazon owl butterflies (*Caligo* sp.), characterized by the huge eyespot on their hindwings. One of the butterflies landed on a girl's hand for a few seconds, making her giggle.

As we rowed back, waving goodbye to Eli and his fellow Bora, their figures gradually disappeared from view. A distant but clear pounding of drums echoed, signaling to the rest of the Bora that the visitors had now left their town.

And as I carefully balanced my way to the front of the boat, with Eli's poignant words and the lingering drumbeats ringing in my thoughts, I wondered about those who had come before the Bora.

A century after Francisco de Orellana's daring venture into the Amazon—marked by tales of complex civilizations and warrior women—other explorers followed, eager to map the region and harvest its abundant resources. However, the Amazon they encountered hardly resembled de Orellana's descriptions. This discrepancy led to the dismissal of his earlier accounts, which were written off as imaginative exaggerations and hallucinations caused by physical exhaustion.

Consequently, the Amazon prior to the arrival of Europeans was often described as an untouched, vast forest with tiny villages that never developed into complex and sophisticated civilizations such as the renowned Incas or the grandiose Maya in Mexico. However, the Incas, known for their large empire in South America, are remembered as extraordinary engineers who built elaborate networks of roads and bridges across the Andes. They were ingenious architects responsible for the awe-inspiring Machu Picchu, and advanced agronomists who innovated agricultural terraces to adapt to the region's challenging topography. Meanwhile, the Maya civilization, one of the most notable civilizations in Mesoamerica, was distinguished by its pioneering advances in astronomy and mathematics, including the development of a complex calendar system and the early use of zero.[4] Their architectural prowess is evident in the impressive pyramid-temples and palaces found in cities like Chichen Itza and Tikal.

The common portrayal of the Amazon Rainforest is that of an impenetrable wilderness, sprinkled with simple, small communities of nomads and gatherers with rudimentary developments and agricultural practices. However, this narrative falls short of the full truth.

Over the last decade, new evidence has emerged that challenges these assumptions. For instance, during the region's worst drought in over a century, ancient stone carvings or petroglyphs, dating back as far as 2,000 years ago, were uncovered in Brazil in 2023. These carvings, depicting a variety of human faces, as well as animals and other natural forms, were sculpted into rocks long before many believed the Amazon had been populated. Archaeologists have suggested that these rocks served as sites for tool preparation, on which inhabitants likely sharpened their tools and arrows. Fragments of ancient ceramics were also found nearby. These findings predate the arrival of Europeans in the Amazon, and add further evidence to the existence of large and complex indigenous villages within these dense forests in pre-Columbian times. In fact, signs even suggest the possibility of older human activities, as the Americas' oldest known rock art, located in the Colombian Amazon, dates back to over 12,000 years ago.

It is this cultural complexity that I find when speaking to the Bora, learning about their profound intellect, mastery of natural resources, and history of intricate traditions.

After my encounter with the Bora, I embarked on a mission to immerse myself in ancient European chronicles of the Amazon, and to seek insights from *Apus* and experts willing to share their wisdom. The more I read, the more I asked, the deeper I wanted to dive in. Over the next year, this curiosity led me to the offices of many outstanding anthropologists, including Carlos, a well-established Andean-Amazonian anthropologist.

– The Amazon was occupied by millions, there is no question about that – he told me assertively.

Carlos is a Peruvian man with a round face and black hair, whose insights seemed as expansive as the Amazon itself.

– A multitude of groups and cultural diversity flourished in the Amazon. Considering the Amazon is almost as large as continental Europe, it only makes sense! he continued. *On one end, there is evidence of very small living spaces with mostly nomadic populations... However, there is also irrefutable evidence that other Amazonian groups built highly developed*

societies with settlements spanning over one hundred hectares. They established cities crowned with twenty-meter-high pyramids, surrounded by highly effective water canals and reservoirs.

As he spoke, he used his hands to illustrate the imposing height of the pyramids, and to trace in the air the intricate networks of aqueducts that surrounded them.

– Pyramids the size of eight-story buildings . . . in the Amazon!

Carlos was alluding to the groundbreaking discoveries of ancient earthen buildings and pyramids in the Bolivian Amazon, estimated to be around 2,000 years old, which were uncovered with the aid of LiDAR (Light Detection and Ranging) technology. This remote sensing technique enables the study of the Earth's surface (and even the moon's) by releasing light pulses from a helicopter, which, combined with other data, produce a three-dimensional representation of the Earth and its surface characteristics.[5] In a place as impenetrable as the Amazon, LiDAR proves invaluable for mapping the forest floor, typically hidden beneath a dense canopy that appears from above as an unending sea of green.

The settlements and structures revealed through this technology indicated that the volume of earth moved to create these Amazonian cities was ten times as much as was shifted during the construction efforts for Akapana, the largest edifice of Tiwanaku, considered one of the most significant Andean civilizations. Notably, Amazonian structures were predominantly made of mud. Unlike the Incas and the Maya, who built with stone, Amazonians rarely had access to stone, and had to resort to the laborious task of molding mud for their structures. This comparison in size and material hints at the Amazonian ancestral civilizations' scale and complexity, suggesting they far exceeded our wildest imagination.

Data released in early 2024 not only supported but also further underscored historical revelations about the complexity and scale of ancient Amazonian societies. The most striking discovery to date was found in Ecuador, where thousands of earthen mounds were uncovered—comprising residential and ceremonial structures arranged in geometric patterns. This marked the largest ancient Amazonian settlement ever found, with an estimated population of 30,000 inhabitants—akin to that of Roman-era London, for comparison purposes. The mounds were interconnected by a network of roads and plazas, all nestled within landscapes shaped by river drainages and agricultural endeavors. This arrangement points to urban centers with clear hierarchical structures and intricate labor systems, indicative of sophisticated, region-wide networks within Amazonian societies.

Furthermore, these studies revealed that the inhabitants masterfully converted a seasonally flooded Amazonian savannah into a fertile agricultural and aquacultural landscape. They constructed complex water-management systems, including artificial lakes, with a primary focus on maize cultivation, alongside hunting and fishing practices, to sustain their thousands of people.

This dexterity in agriculture is not to be taken lightly.

There has long been a misunderstanding that the Amazon couldn't harbor great civilizations because of its soil. Tropical soils like those in the Amazon are notorious for being poor, thin, and slightly acidic. Some areas even feature white, sandy soils: the result of erosion spanning hundreds of millions of years, stripped of fertility and essential minerals, and thus unsuitable for crop cultivation. The theory goes that if you can't feed people, then society can't develop.

However, radiocarbon dating has shown that natural wildfires and the seasonal flooding of rivers created patches of highly fertile land in the Amazon over millennia, well before human settlements took root. Archaeological findings reveal that indigenous peoples had learned to identify and use these unique fertile spots—rich in calcium, phosphate, and charcoal—long before the formal invention of agriculture.

Archaeological discoveries now verify the Amazon's role as a pioneering hub for plant cultivation and soil management, marking it as one of the earliest regions on the planet where papaya, peanuts, tobacco, cacao, and more were grown.

As Carlos sifted through his library, looking for a book to show me undeniable proof of an important aspect overlooked by many early European chronicles, I remembered a story my father had once shared with me about Amazonians' mastery of water and soil use.

During his early years as an engineer, my father traversed the Andes and the Amazon, working with communities in remote areas to develop sustainable technologies inspired by ancestral techniques for harvesting water from the air and ground. In Central Amazonia, he encountered an indigenous community that used to consume water directly from springs and rivers, just as their ancestors had done.[6] However, prolonged droughts had caused water levels to plummet, rendering these vital sources of potable water nearly inaccessible.

Using drilling rigs repurposed from oil extraction, my father and his team managed to create a well about 5 meters (16 feet) deep, tapping in to a source of pure, clean water. Although many factors need to be considered (including potential soil erosion), this initial achievement hinted at the possibility of constructing underground water channels in the Amazon. These channels could potentially provide drinking water to communities residing deeper within the rainforest, in areas less served by constantly flowing rivers. As innovative as it was at the time, they eventually learned that other remote Amazonian communities had already been using a similar technique to access water when the rivers were low, and the rain didn't come.

What's remarkable is that the archaeological discoveries of earthen cities in the Amazon have brought to light the fact that our Amazonian ancestors had exquisite water control and distribution systems. These sophisticated channels and drainages delivered water to areas distant from rivers and lakes. Water reservoirs were also found, suggesting that previous civilizations might have faced droughts in the rainforest and devised creative solutions to overcome these challenges.

– *Here it is!* exclaimed Carlos; he had finally located a copy of the book he'd written, which he was keen to share with me.

He directed my attention to a photo on the top of the page. I squinted, trying to balance the page's bright whiteness against the photo's darker hues. The image showed large rock constructions that resembled the remains of a building or tunnel. The entrance was framed by long, rectangular rocks, standing 4–6 meters (13–20 feet) tall.

The rocks used in this construction aren't naturally found in the Amazon, indicating they must have been shaped elsewhere and painstakingly transported to the rainforest. In and around the building blocks, long green plant stems and leaves were growing, adding a picturesque quality to this extraordinary finding, nestled amid infinitely tall trees and a dense forest floor. It was a resilient structure that had survived the changing conditions of the rainforest, including the exuberant vegetation and the continual flooding.

I had seen photos of the mud constructions scattered across Bolivia, Ecuador, and Brazil, but nothing made with stones erected in the core of the Amazon. Upon close examination, I noticed the rocks showcased architectural methods similar to the ancestral *pirqa* technique, a stone-building practice that originated in societies pre-Inca and was later refined by the Inca Empire. But the Incas didn't enter the Amazon . . .

– *Rock buildings? In the Amazon?* I was in a state of shock. I was trying to find the words to formulate what I would say if I stumbled upon such a structure while trekking through the jungle. I would probably suspect I was severely dehydrated.

– *The European chronicles missed another important thing* – Carlos declared proudly, flipping through more photos depicting similar rock constructions peppered throughout the Amazon. *I've read all the original texts, and many confidently stated that the Incas didn't penetrate the rainforest—that the Amazon was the only area the Incas couldn't conquer in South America, even with their infinite gold and supreme power.*

His words echoed the articles I'd read, writings that declared that every attempt by the Inca—rooted in the high Andes of Cuzco in Perú—to subdue the Amazon had ended in defeat, their quest to dominate the rainforest proving futile. Just as it had the European travelers, these articles claimed, the Amazon Rainforest had consumed them instead.

– *But these findings prove otherwise. They virtually change the map of the* Tawantinsuyu *(Inca Empire)* – Carlos said.

He went on to explain that these rock constructions suggest that the Inca did advance into the Amazon, developing outposts, possibly to support trade routes. Many of these remains have been discovered, not just in isolated locations but also across the expanse of the Amazon in Brazil, Bolivia, and Perú, and even further south in Argentina, slowly enabling us to reconstruct a more accurate depiction of the past.

– *It changes our understanding of the Incas and the Amazon—and it makes sense!*

His excitement was palpable, and in his fervor he pre-empted my questions.

– *These findings show that the trading network between the Incas and Amazonians was so much broader than we thought. How else did the Incas have access to exotic Amazonian bird feathers or spices? How did Amazonians have constant access to salt for consumption and food preservation beyond* collpas?

Collpas, natural salt licks also known as *saladeros*, are located in forest clearings along riverbanks, or in mountains where mineral-rich springs start. These natural mineral deposits are essential for animals, like parrots and macaws, capybaras, tapirs, and others, which come to lick essential minerals needed in their diet to ensure health or counteract toxins. It is not practical or sustainable to extract salt for human consumption from these deposits, but if you ever want to photograph dozens of macaws in one place, look for these salt licks. The resulting photos are outstanding.

Carlos's book and his insightful explanations add yet another layer to our understanding of the complexity of Amazonian civilizations: their skill at harnessing natural resources in surplus for trading, extending their reach to enable them to procure goods not inherently found in the rainforest. With trading networks between Amazonians and the Incas as expansive as the evidence suggests, it is possible to imagine that Amazonians might have received gold and silver in exchange for their unique tropical goods. Ancient chronicles suggest that remote cities in the Amazon were filled with gold and silver, birthing one of the most enduring legends and myths in the Americas: the lost city of "El Dorado" (the golden one). This was a legendary city thought to be rich in precious metals and situated somewhere in the unexplored corners of South America, potentially the Amazon. Another related legend, Paititi, speaks of a hidden Inca city believed to lie east of the Andes, said to hold vast treasures left behind by the fleeing Incas. Both legends drove numerous expeditions to the region, impacting exploration and colonization, and even inspired famous animated movies.

– *How have I never heard of this before, Carlos?* I asked, marveling.

Carlos laughed, his eyebrows arching in genuine amusement as he handed me the book to keep. In his realm, this information was common knowledge, yet in mine—and likely for most—it was a new world to discover.

With a newfound love for archaeology and history, I thanked Carlos for his valuable time as he got ready to teach an upcoming class. Many theories and questions buzzed in my mind, and I rummaged through my purse for my phone to jot them down, looking to make sense of my thoughts. I wondered about the potential insights that genetic and microbial studies on these archaeological sites might reveal—whether they be about ancient plant-based therapies and diseases, or the historical lineage of medicinal honey use. What other questions remain unasked, lurking in the shadows of our understanding, awaiting discovery?

The desire to have ownership over the Amazon has never been successful. In reality, the Amazon Rainforest offers herself to us, revealing only what is meant to be seen and heard at the time, and holding onto secrets and stories that are now at risk of being lost, as she fades away between our fingers. We must hope her inhabitants can remember the old ways.

The archaeological findings and stories of trading also point toward something else: the constant exchange of culture between the peoples of the Amazon and the Andes, laying the foundation for the widespread reverence of the Pachamama.

Culture, like nature, extends beyond political borders, flourishing where nourished, giving life to new knowledge, stories, and hope—just like the mountainous and serpentine Andes connect to the luscious and humid Amazon in multitudes of ways, and always have, since their origins. Through the sharing of stories and philosophies—such as the art of "living beautifully" (see page 15) and the meaning of Pachamama (see page 285), culture shapes our past, present, and future. Thus, the protection of the culture, languages, and stories of indigenous communities becomes an essential right, holding as much importance as the conservation of biodiversity itself.

It is true that the Amazon is a natural paradise of biodiversity, harboring at least 9 per cent of the world's known mammals, 14 per cent of birds, 8 per cent of amphibians, 13 per cent of freshwater fish, and 10 per cent of known plants. But its significance goes beyond these statistics. The Amazon is a place of biocultural heritage where the indigenous cosmovision does not divide nature from culture, but rather teaches us profound lessons on sustainable living. Learning from ancient civilizations that emphasized respect for their natural surroundings means creating pathways for societies to flourish in harmony with the Pachamama. We can find inspiration from ancestral cultures that stand as testaments to the possibility of crafting realities where both biodiversity and people can thrive equally. Perhaps this is the power of preserving culture: the possibility of remembering how to live in synchronicity with our planet.

11

IN THE EYES OF THE JAGUAR

Dragon's blood

SCIENTIFIC NAME: *Croton lechleri*

TRADITIONAL NAME: *Sangre de grado* (Dragon's blood)

ORIGIN: South America's tropical rainforests, including the Amazon

TRADITIONAL USES: This tree has a long-standing history in traditional medicine, with the sap, also known as "Dragon's Blood," revered as a powerful cure for a wide range of conditions, from hemorrhoids to incontinence and various skin conditions. Locally known as "liquid bandage," the sap is applied to wounds to stop bleeding, prevent infection, and accelerate healing. In spiritual practices, it is believed to ward off negative energy. In the 17th century, Bernabé Coco, a Spanish naturalist, was one of the first to document the use of this ancestral medicine in the written form, noting that the sap was used as a "brew that stops bleeding."

SCIENTIFIC INFORMATION: *Croton lechleri* is a tall, flowering tree, 10–20 meters (30–65 feet) in height, with a very narrow trunk. It is easily identified by the dark red sap that oozes from the tree when the bark is cut. This sap contains a variety of chemical compounds, including taspine, which has been shown to help with closing wounds. It is also rich in proanthocyanidin oligomers, which are powerful antioxidants. These oligomers contribute to the unique properties of crofelemer, one of just two botanical drugs ever approved by the FDA. This pharmaceutical is used to treat HIV-associated diarrhea. Additionally, the sap exhibits high levels of antimicrobial activity, and is used internally for treating various gastrointestinal disorders, such as gastritis, ulcers, and infections, due to its anti-inflammatory and antiviral properties. Responsible harvesting is necessary to ensuring the survival of this ancient tree and safeguarding its role in traditional and modern medicine.

Something was different. I was still breathing, but it felt as if something was pressing against my nasal bone. It wasn't as if my breathing was restricted, but rather as if the movement of oxygen was being redirected. My nose was twitching, and my upper lip was slightly open, as if I was trying to inhale more than just air. It felt as if my whole face was changing shape, with the nose becoming the most reliable of the senses that was central to my being.

I jumped gracefully, landing atop cold grass, my four limbs evenly distributing my weight on the ground. As I landed, my shoulder blades were pulled back toward my spine, helping absorb the impact. My shoulders began to move slowly, rotating slightly in their sockets, gliding in a synchronized fashion with the movements of my limbs. It felt as if I was walking on high heels, the feminine motion extending throughout my body toward my hands and fingers. Except I wasn't.

Looking down, I realized that instead of human legs and arms, I had muscular limbs and rounded, large paws with thick pads to silence my movement. My body was covered in a dense, sleek black fur, with a long and majestic tail seamlessly blending with the rest of my coat. In the deep ayahuasca dream, I had become a black jaguar.

Tapping my index finger and thumb together, I reassured myself that I was still in control of my body and aware of my physical surroundings. I was lying on a mat on the cold floor of an enclosed room, with my husband beside me on my right. Across from us, the shaman sat cross-legged, with her eyes closed in deep concentration.

The experience of transfiguring into an animal within a dream was utterly new to me. The sensorial and emotional effect was so vivid and realistic that my heart was beating rapidly. In the background, I heard the shaman's melodic singing, the *icaros* (see page 129) inducing a sense of calm and comfort, reassuring me and giving me the confidence to breathe deeply and return to the dream.

I had never been so aware of the thousands of smells that surround us. It was as if I had gained the ability to take hundreds of quick inhalations within a few seconds. I could recognize everything around me simply by smell, identifying traces that provided a clue as to what was here before and what may lie ahead. It was like my nose became my eyes into the past and into the future. With my jaguar body, I kept moving across the foliage, attuned to every vibrant pulse of the rainforest. I felt powerful.

In the dream, beneath my paws, I sensed the earth transitioning to a soft and wet forest floor as I neared a river. I could taste the freshness of the river's water in the air; it carried a multitude of scents from distant lands and countless creatures that merged together like the distinct herbs and spices that compose Amazonian soups in Perú. I followed a path running several meters parallel to the river, and found some fallen trees that had created a secluded and hollow space, providing natural shelter. My tail elongated and gently swayed behind me, helping to counterbalance my weight and signaling my calm alertness. It felt almost like a fifth limb, granting me the confidence to make subtle adjustments in position and movement, essential for maintaining my stealth.

As a black jaguar, I was on a mission to find a new a home for my cubs. Our previous den had been ravaged with illness, sickening my cubs and forcing me to seek out a safer, healthier home. With this particular focus, I moved through the rainforest.

I traveled silently around the dense underbrush, my body blending seamlessly into the shadows. My black fur absorbed light rather than reflecting it, rendering me nearly invisible in the dappled sunlight that filtered through the canopy. Every step was calculated, with my amber eyes scanning and my nose drawing in the scents of the moist, crisp flora, and the underlying traces of nearby prey. At the furthest corner, I detected the potent aroma of medicinal plants and flowers that could help heal my cubs.

My maternal instincts were magnified as I searched for a safe den site. Somewhere deep inside, the area felt right. It had everything we needed, from ample food sources to natural remedies for my litter. My sense of smell never failed, dissecting the rainforest down to the grain, ensuring no detail was missed. I lingered a while to detect any signs of intrusion or danger, marking my new territory and signaling to others that this area was now occupied.

I knew this was the place. It was not just about survival; it was about providing a future for my cubs.

The entire time I was a black jaguar in the ayahuasca dream, three abilities dominated my body: instinct, confidence, and presence—the ability to be fully focused on the moment. This trifecta of superpowers allowed me to navigate the rainforest with discerning eyes (and nose), knowing with certainty where to explore next for the well-being of my species. It is a powerful metaphor for the nature conservation and regeneration goals I aspire to achieve with my work.

The morning after the ceremony, three insights came to the surface as I spoke with Inin Kena, the female Amazonian shaman who had led the ceremony.

– *You have a powerful connection to the animal world. Its wisdom is speaking to you, Rosa* – Inin Kena explained, indicating that many seek a deeper connection through ayahuasca to understand the jaguar's world.

And your intuition is very strong – she added, her black eyes piercing mine. She tapped her stomach, advising me to always trust my gut. I resisted twitching my nose.

– Maestra *(master), in the dream, I could also detect medicinal plants solely based on their smell; I knew which ones were therapeutic and would be useful to cure my cubs* – I told her, fascinated by the idea that jaguars may self-medicate.

Inin Kena simply smiled and nodded, her eyes glistening with a knowing wisdom, as if she held onto secrets that she well understood, yet must patiently wait for others to uncover when they were ready.

Inin Kena is one of the wisest and most interesting people I've ever met—not just because of her rare status as one of the very few female shamans in South America, but also due to her remarkable heritage and life story of resilience. Historically, women's involvement in ayahuasca ceremonies was limited to the preparation of the sacred brews, while men, deemed the true medicine administrators, participated in its consumption. This practice stemmed from various traditional beliefs, including menstruation's compatibility with the plants, and the intense demands of shaman training, which include more than ten years of abstaining from sexual activity, meat, salt, sugar, and alcohol—making childbearing a significant obstacle. Yet, in recent years, courageous women like Inin Kena have lifted their voices and challenged these norms, pursuing the rigorous path of shamanic training with unwavering determination and commitment.

What I personally found beautiful was the feminine energy that Inin Kena radiated and transmitted throughout the ceremony: a maternal voice guiding the dreams, providing a safe space for discovery. I like to imagine that this emotional safety was the fertile ground that allowed my unconscious to dive deep into ancestral knowledge, reaching experiences and information I had never even dreamed of. It awakened a maternal instinct that intertwined with my scientific pursuits to protect the natural world, reminding me to trust my intuition when seeking answers or solutions in the Amazon. It was a reminder that my indigenous upbringing does not dilute my scientific training, but rather enhances it, fostering unique ideas that contribute to preserving our rainforest and wildlife. Perhaps it is a feminine voice that could amplify the fears and hopes of the rainforest, inspiring more respectful explorations of the Amazon aligned with indigenous worldviews.

The name "jaguar" originates from the South American indigenous word *yaguara*, which means "animal that kills with one leap." The jaguar, scientifically known as *Panthera onca*, is the only species of its kind in the Amazon Rainforest. The most commonly observed jaguars bear a golden-brown coat with a pattern of circular dark spots known as "rosettes." However, some jaguars exhibit a genetic variation known as melanism, which results in a predominantly black coat. Although these melanistic jaguars, often referred to as "black panthers," appear black, their spots are usually visible under direct sunlight or upon closer inspection. Currently, about 6 per cent of all living jaguars are melanistic. These darker jaguars are believed to be more commonly found in denser and wetter parts of the rainforest, where their black fur provides them with better camouflage.

Jaguars are opportunistic hunters. Whether by night or day, they prey on almost anything they find, from capybaras, armadillos, tapirs, monkeys, and birds to iguanas, fish, and caiman. Relative to their size and weight, jaguars have the most powerful bite in the feline world,[1] with their teeth strong enough to pierce through skulls. Female jaguars usually give birth to two or four jaguar cubs, with their pregnancy lasting around 14 weeks. Jaguar cubs are born blind and weigh about the same as a loaf of bread, so they are totally dependent on the mother for protection and nourishment. Their eyes typically open within two weeks of birth, and they nurse for a few months. Around the time they turn six months old, cubs start to accompany their mother on hunting expeditions to learn the skills necessary to survive. But within two years, the cubs approach full adult size and become independent, finding their own territory, which reduces competition for natural resources and increases genetic diversity, ensuring species preservation.

In ancient times, the jaguar, also known as *puma* or *otorongo*, became a divine animal that represented the world of the living.[2] This meant that the jaguar was not only revered as a powerful animal but also as a spiritual bridge between the earthly and the divine realms, embodying life force and vitality. This profound connection was seen at the heart of spiritual and cultural life across the Americas, from North America to Argentina.

The Inca, the Aztec, and the Maya worshipped jaguars as gods. They carved the animal's shape and image into temples, thrones, and even funeral shrouds, immortalizing their role in both life and the afterlife. In Perú, locals associated the roar of thunder with the jaguar unleashing its fury, while in Colombia, communities believed that the jaguar was manifested in the sun, linking it directly with celestial power.

This deep reverence extended into multiple cultural expressions. Throughout pre-Columbian America, jaguars were brought to life in ceramics, pottery, textiles, and creations in gold. Artifacts depicting jaguars have been found across the Andes and the Amazon, indicating their ubiquitous cultural significance. Some Amazonian cultures even hunted jaguars to wear their fur, drink their blood, and eat their hearts in rituals aiming to infuse participants with the jaguar's strength.

Today, the legacy of the jaguar's divine status continues to influence local traditions. Several communities wear colorful jaguar masks and spotted costumes in parades and festivals, like the Amazon Carnaval or the Tigrada Festival in Mexico, where the jaguar god Tepeyollotl is praised to bring about abundance and rain. This cultural and spiritual relationship with jaguars even extends into the commercial world. For example, the popular Amazonian beer San Juan proudly displays the image of a roaring jaguar.

As a symbol of the power and strength of nature, jaguars have given rise to multiple myths, including the tale of the *Yanapuma*, which in Quechua translates to "black jaguar." This mythological creature, revered and feared in the Amazon, is said to have fur as black as the night itself. Among the various versions of the legend, some depict the *Yanapuma* as a formidable being with the unique ability to transform into a man during the full moon. It is widely believed that the *Yanapuma*, due to its cunning and stealth, can never be seen by the human eye, making it the most ferocious predator in the Amazon.

Other tales suggest that certain individuals, often regarded as wizards, possess the mystical power to morph into jaguars. This transformation allows them to blend seamlessly with the dense rainforest, and only a select few have the capacity to take on the form of the elusive black jaguar.

During sacred ayahuasca ceremonies, the spirit of the jaguar, including the *Yanapuma*, plays a crucial role. Some of the most powerful shamans summon this spirit to harness its strength and protection.

They believe that the jaguar guards the ceremony against dark forces, catalyzing transformation and expelling disease. There are also those who describe a spiritual journey where they metamorphose into jaguars themselves. This profound transformation allows them to visit the lands of their ancestors. In these lands, they encounter powerful spirits and grandfathers who impart ancient wisdom, reveal secrets of medicinal plants, and offer loyal guardianship.

The dream where I became a jaguar led me down a rabbit hole of reading that forever impacted my approach to scientific inquiry and integration of indigenous wisdom. This exploration brought me to the study of "zoopharmacognosy," a fascinating field that examines how animals use medicinal plants to treat their own ailments. It suggests that humans might have first learned to use medicinal plants by observing animals, who have been self-medicating since long before humans documented these practices. World-renowned biologists have previously mentioned that drinking alcohol wasn't even invented by humans; chimps, elephants, and other animals have probably been getting drunk on fermented fruit since the development of their species.

Animal self-medication is more widespread than most people are aware of. In the early 1980s, scientists observed chimpanzees in Africa carefully selecting and eating the leaves from a tropical daisy called *Aspilia*. The chimps swallowed the leaves whole, without chewing. Laboratory analyses of their feces later revealed the remains of the leaves, along with expelled intestinal worms, suggesting a medicinal use: the rough texture of the leaves acted like "Velcro," helping to purge the parasites from their systems.

Intriguingly, local people also use the leaves and roots of *Aspilia* to treat cuts and infections, and they have been discovered to contain a variety of bioactive compounds, including anti-inflammatory flavonoids and antimicrobial agents. This discovery led ethnobotanists to wonder whether the observation of animals using plants for health purposes could have encouraged humans to explore various parts of the same plant for medicinal reasons, expanding ethnobotanical knowledge.

In the following years, similar observations were noted between chimps and other medicinal plants like the *Vernonia* bush. Chimps infected with nematodes (roundworms) ate the plant, significantly reducing their nematode-egg-per-gram infection, quickly recovering energy to resume hunting, and only consuming the plant again in periods when parasite infections were high. The local people claimed to consume the same plant to treat malaria and dysentery. The medicinal properties of the plant were then tested and verified, leading to the identification of molecules with anti-worm, anti-amoebic, and antibiotic properties.

Other large animals have been observed self-medicating with a neighborhood pharmacy, such as bears, deer, elk, macaws, and lemurs. A pregnant elephant was observed devouring an entire tree of the borage family, before giving birth a few days later. Curiously, pregnant women in Kenya were found to consume tea made with the bark and leaves of the same tree species in order to induce labor. Moreover, woolly spider monkeys in Brazil were found to control parasite populations with the same plants that are used by Amazonian people to control parasite infection and related diarrhea in traditional medicine. The monkeys have also been observed consuming plants that modulate their fertility, with some leaves serving as birth control after they have given birth while other plants enhance their chance of pregnancy. These practices and observations might have encouraged local communities to integrate plant-based birth-control methods, further expanding the knowledge of plant medicine in human culture.

Interestingly, self-medication behavior is not restricted to animals with high cognitive abilities. Instead, it is postulated as a type of adaptive plasticity: the ability to adapt behavior, genomics, and biological processes in response to the environment throughout lifetime in order to improve chances for survival and reproduction. So it is unclear how much of these behaviors are due to knowing versus learning. The list of animal pharmacists continues to grow and now includes caterpillars, butterflies, bees, lizards, and fruit flies. In the case of lizards, they eat a specific root to counter the venom of a snake when bitten. However, these behaviors aren't solely restricted to treating disease; they are also used as preventatives.

In my own research, I've observed a remarkable behavior in Amazonian stingless bees (whom we met back in Chapter 5): they harvest the deep-red resin of dragon's blood trees to construct their hives and honeypots, making them less prone to disease and pests. We captured the behavior with macro photography, which showed that the bees use "pockets" within their legs, typically reserved for carrying pollen, to transport the resin back to their nests. It is hypothesized that when this antimicrobial resin is mixed with the bees' wax, it creates a unique building material that is often referred to as "bee glue,"[3] or more formally known as propolis. This combination likely plays a crucial role in fortifying homes against harmful microorganisms and invasive ants, thus enhancing the bees' health and quality of honey.

Other animals ingest specific natural compounds in order to deliver poison or deter predators from eating them, rather than to cure disease.

This is how dart poison frogs become toxic; they consume insects, such as beetles or ants, that are highly poisonous, and store the molecules in their skin, ready to secrete when they need to defend themselves. When raised in captivity, the amphibians fail to produce the same toxins.

Interestingly, Neanderthals—who are part of the hominin group that includes our direct ancestors and species more closely related to us than to other great apes—also used medicinal plants. Studying the DNA of dental plaque on the fossilized teeth of Neanderthals revealed that these hominins consumed plant leaves rich in salicylic acid (the active component in aspirin) and penicillin when infected with *Enterocytozoon bieneusi*, a microbe that causes gastrointestinal issues including severe diarrhea. This suggests that Neanderthals possessed great knowledge of medicinal plants, potentially self-medicating even 40,000 years before penicillin was formally discovered as a medicine.

In the case of jaguars, little research has been done on their use of medicinal plant life. Jaguars in the Amazon have been observed gnawing on the roots, bark, and leaves of *B. caapi*, the main active ingredient in ayahuasca, before rolling on the ground. Some have suggested that this behavior is similar to domestic cats nibbling on catnip to achieve a state of pleasure and relaxation. Some Amazonians even credit their use of ayahuasca to observing jaguars consuming the plant. But it has not been verified whether jaguars consume the vine when ill, or to potentially dilate their pupils, enhancing night vision, which is a common effect of the plant. It could be that further exploration into the zoopharmacognosy of jaguars leads to the discovery of new plant-based medicines that could benefit both humans and animals. These remedies could help us combat the growing epidemic of multidrug-resistant infections and other health conditions that climate change is exacerbating. This possibility underscores a profound truth: there is so much left to learn from the natural world and the untapped wisdom animals hold.

The first face-to-face encounter I had with an Amazonian wild cat was when I was five.

I was playing with Barbies with my cousins upstairs when we heard the sounds of someone arriving at our house. A distant uncle had just arrived from the Amazon bearing a gift: a tiny kitten with small, rounded ears and golden-brown fur adorned with dark, irregular stripes. When we came running down the stairs, I saw the kitten cradled in my uncle's hands, purring, with large expressive eyes that looked like gems. I jumped with excitement, feeling eager to get to know my new friend.

During its first weeks with us, the precious kitten was very loving. I remember looking forward to petting him every time I came back from school. However, soon enough, he began springing with agile grace from wall to wall, snagging and ripping my mother's curtains with his retractable claws. His playfulness quickly turned dangerous as our living room became his jungle gym.

A run to the vet confirmed that this was no exotic-looking domesticated cat, as we'd all genuinely believed—my uncle included. Instead, our kitten was a *tigrillo* (*Leopardus* sp.), a small, primarily nocturnal and solitary wild cat found throughout the Amazon Rainforest. Known as *oncillas*, these cats have an elusive nature and excellent climbing skills, and primarily feed on small mammals, lizards, birds, and invertebrates. Although *tigrillos* tend to avoid human contact, they can become defensive if threatened or cornered. Within a day of the vet's verdict, my family sent the *tigrillo* to a nature recovery center in the Amazon to ensure he learned to be independent before returning to the wild. Seeing the *tigrillo*'s behavior up close prompted me to inquire about the stamina and strength of larger wild cats.

In all my years working in and exploring the Amazon, I've never come across a jaguar—that I know of. Unfortunately, in a way, this reflects a reality that is uncomfortable and unavoidable: jaguars in the Amazon are severely endangered.[4] Habitat fragmentation from high rates of deforestation forces jaguars into smaller areas of the forest, meaning they are not able to travel far to mate. This leads to inbreeding, loss of genetic diversity, and local extinction. Poaching is another major threat, as jaguars are hunted for illegal trading of their teeth, bone products, skin, and claws. They are also captured and sold as exotic pets.

However, jaguar encounters in the Amazon do still occur. In some instances, these encounters are peaceful. Chances are the jaguar will look at you for a few seconds and then simply continue its journey. Avoid sudden movements, and slowly back away without turning your back on the animal or running away, and you will live to tell the tale. This has been the experience of various colleagues, including some who have encountered female jaguars bearing cubs while standing a few meters away from them, or when a male jaguar came sniffing around outside their camping tent, curious about the people's doings.

But not everyone's that lucky.

A few weeks before my jaguar ayahuasca dream, I was trekking through the rainforest with our team and Manuel, a trusted indigenous partner who had been with us on previous expeditions. We had been trekking for over an hour, and as we advanced further, Manuel shared stories of his previous animal encounters, including those with

Amazonian "cats," as locals call any feline including *tigrillos*, ocelots, and jaguars.

In his early 30s, Manuel had delved into unspoiled and remote corners of the Northern Peruvian Amazon, reaching depths he had never explored before. He navigated the jungle for hours on end in search of a new catch, either a *majás*[5] (an herbivorous rodent) to feed his family or a snake to sell in the local market to buy some food. He clung to his trusty companions—garlic, salt, and Amazonian tobacco—in fear of disrespecting or disturbing the evil spirit of the *chullachaki*.

However, on that particular night, it seemed as if most spirits were in slumber, along with the mammals and snakes that Manuel was after. Disappointed, he began to retrace his steps, already dreaming about the hearty rice soup that would await him upon returning home.

Suddenly, he came face to face with a cat.

The cat's imposing eyes reflected the moonglow, piercing through the thick vegetation like unforgiving X-rays in the midst of pitch-black Amazonia. Its whiskers twitched as if, unlike Manuel, it had just found what it was seeking that night. The rainforest fell into an eerie silence, broken only by Manuel's heavy breathing.

Based on Manuel's descriptions, it was a *Panthera onca*—a jaguar. Only a few ancient trees and scented bushes, about 10 meters (30 feet) away, separated Manuel from one of the most ferocious predators in the entire Amazon.

Instinctively, Manuel looked down, trying to make himself as inconspicuous as possible, hoping the cat wouldn't see him as a threat. He wished, as intensely as he had ever wished for anything in his life, for the jaguar to find him inoffensive and to simply walk away. However, the jaguar had different plans.

With enviable elegance, it shook its head and took a deep sniff of the air before showing its teeth. It looked directly into Manuel's eyes, as if stripping away his thoughts and dreams, inviting him to witness his entire life flashing before him. Its gaze was so intense that it caused Manuel's eyes to glisten, though he couldn't recall if this was due to his pupils adjusting to the newfound fear in the profound darkness, or the fact that tears were welling in his eyes.

– I just thought about my kids and my wife. I'd left home in such a rush, not even hugging them goodbye. I knew they would be waiting for me to return in time for breakfast. They would worry if I was late.

Manuel prayed to all the spirits and *Apus* he knew, promising to bring *ofrendas* to the *chullachaki* if the rainforest spared his life. He silently whispered that he didn't want to become another unidentified body, unrecognizable and eviscerated in mere moments, a victim of nature's precise attack.

He summoned the last few ounces of courage left in his muscles and scanned the area for any trees he could climb. Unfortunately, there were none tall enough to hide him from the jaguar, a species known for its tree-climbing prowess. His chest constricted further as panic set in.

Part of him still had faith that the cat would just leave. But it didn't.

Instead, it began to move closer, actively growling at Manuel. In a desperate attempt to save his own life, Manuel drew the pistol out of his pocket and pointed it at the animal. He couldn't distinguish the tears streaming down his face from the heavy rain that had started pouring a few minutes earlier. The jaguar displayed its teeth one more time, jumping forward and charging at Manuel.

One shot.

Two shots.

Three shots.

Manuel fired frantically, begging the rainforest—and the jaguar—for forgiveness. To his horror, the jaguar was still standing, and terrifyingly closer than before. Manuel panicked, unable to comprehend why the jaguar was still on its feet, only to realize that the bullets had fallen to the ground next to him, as lifeless as dead flies. Manuel continued to fire, each shot a desperate plea for survival. He must have fired 15 times that evening, with all the bullets accumulating at his feet. It might have been the *chullachaki* sparing his life, accepting the Amazonian tobacco he carried with him that night, or perhaps it was a final stroke of luck bestowed upon him, but on the 16th shot, the bullet found its mark, striking the jaguar straight in the heart. With a sound like a tree crashing to the forest floor, the giant cat's lifeless body collapsed.

– I cried and cried as I ran back home – Manuel told me, his voice breaking now as we kneeled to crawl under a fallen log.

Manuel narrated his story intently as we navigated the rainforest at night, jumping over lianas, avoiding poisonous snakes, and shooing away giant moths. Electricity ran down my back; I felt like I had just lived his experience. My heart felt heavy with Manuel's lingering pain and the thought of the dead jaguar.

There were many lessons embedded within Manuel's pain that have since influenced my relationship with the animal world. I can only suspect that many wouldn't hesitate for a second in shooting an animal if they were under threat. In fact, some people even hunt animals purely for sport, power, or the alleviation of boredom, without lingering thoughts or remorse.

Instead, Manuel showed reverence. His indigenous perspective holds that animals have spirits residing within them, and when hurt, their spirits suffer too. In Manuel's eyes, he wasn't just defending himself and saving his own life—he was ending the life of another sacred being, sacrificing its flesh and spirit for his own. Manuel believed that he had interrupted the natural cycle of life, and as such, his actions needed to be compensated for. He wouldn't want to disrespect the Pachamama that provided him with a home, food, and a livelihood.

Manuel's reverence for the animal world gave me hope. It reminded me that it is possible for humans to develop a connection with nature so strong that it becomes instinctual: a connection that serves as the foundation to further our knowledge and take action with hopes of preservation.

The animal world is complex, and so is the cultural relationship our ancestors developed with animals—one that I believe still has so much to teach us if we are brave enough to deconstruct existing theories and rebuild our perception of the world. Ancestral ways of living and learning from animals may offer a secret vault of sustainable solutions to what ails our planet—much like the ancient forms of communication in the animal world that inspire a renewed sense of wonder and curiosity in those who dare to listen. This inspiration drove me to embark on my next expedition, on which I discovered the language of light that brightens the darkest corners of the rainforest.

12

LIFE THAT GLOWS IN THE DARK

Matico

SCIENTIFIC NAME: *Piper aduncum*

TRADITIONAL NAME: Matico or *hierba del soldado* (the warrior's herb)

ORIGIN: Native to tropical Central and South America, including the Amazon Rainforest

TRADITIONAL USES: Matico is a highly valued native plant in the Amazon Rainforest. Indigenous communities use the leaves for their antiseptic properties, healing wounds, preventing infection, and stopping bleeding. A tea is also prepared using the matico plant as the base ingredient, and used to treat gastrointestinal issues including ulcers and hemorrhages. It is also used to relieve menstrual pain, given the anti-inflammatory and antiviral properties believed to be found in the plant. Oil made with the plant is also used as an insect repellent. History says that a wounded European soldier named Matico learned from the local communities how to use the leaves of this plant to stop bleeding.

SCIENTIFIC INFORMATION: Matico is a flowering plant that grows to a height of 3–7 meters (10–23 feet), with lanceolate leaves and cord-like spikes of various tiny flowers ranging in color from white to pale yellow. The plant has a spicy or peppery aroma due to the high concentration of flavonoids, terpenes, chromenes, and alkaloid molecules. Scientific research has shown that the plant has broad-spectrum antimicrobial, antiparasitic, and insecticidal properties, and can protect against malaria-carrying mosquitoes. The leaves, used to make essential oil, are rich in camphor, viridiflorol, and piperitone, and are particularly potent with antifungal, antibacterial, and antiviral properties.

THE SPIRIT OF THE RAINFOREST

Deep in the rainforest lies an extraordinary wonder that few know about.

A wonder that may forever change how you view the rainforest, or the Pachamama herself.

To get there, you must complete a series of challenges that are not for the fainthearted: trek into virgin jungle, cross violent rivers, climb up sinking hills, and crawl on the ground beneath fallen tree trunks.

While navigating the rainforest, beware of ferocious jaguars, venomous snakes, and killer spiders the size of birds, not to mention the soul-snatching spirits that roam these shadowed lands.

Press on through sections of the rainforest so dense that not even your headlamp can pierce the darkness.

And if you are brave enough to arrive in the heart of the jungle at night, that's where you will find the extraordinary wonder: trees that glow like bright blue bulbs, an intense glow that symbolizes the spirit of the rainforest.

This tale fueled my imagination when I was only seven. *Trees that glow like bright blue bulbs.* The story sparked endless wonderings and conjectures about their existence in the heart of the Amazon at night. Why would trees glow? Did anything else emit such light? How was it possible? I couldn't understand why nobody else was talking about it, and I was obsessed with uncovering this mystery. My mom wouldn't let me delve into the jungle at night for fear of snakes and jaguars, and so I would seek solace in children's books that narrated similar stories, fantasizing about myths of luminescent life in the pitch darkness of the rainforest, casting a rainbow of blues, greens, yellows, and reds like a Christmas light show.

At nine, the tale became more deeply ingrained in my mind when an indigenous man shared his own story of being in the Central Amazon after dark.

– *The spirit of the glowing tree is powerful* – he warned. *The elderly teach us not to touch it, and to leave it intact as a sign of respect, or else we may face its anger.*

I listened intently, without blinking, wondering if the tree would also be upset if I just had a glimpse of it.

– *I've seen them* – he continued. *They shine blue as the sky. They live very, very far deep into the jungle, as far as you need to walk to see cats that are now becoming harder to find in the wild.*

That night, I went to bed dreaming of enchanted forests that glow in the dark. But as time went on, and I grew into adulthood, I forgot about the tale of trees that glow like bright blue bulbs—that was, until I returned to the rainforest.

We were advancing carefully. Our headlamps pierced through the canopy, attracting dozens of critters everywhere we turned. In my right hand, I held a metal snake grabber and gently tapped the ground in front of me three times with caution; this was my first time trekking through this part of the rainforest, and the wet leaves and vegetation created the perfect nest for poisonous scorpions, spiders, and snakes. We were here to uncover the wildlife at night in one of the most enigmatic, endangered, and least-known ecosystems of the Amazon Rainforest: white-sand forests. These are rare, nutrient-poor ecosystems that host unique specialized biodiversity while facing growing threats from commercial white-sand exploitation and deforestation.

This was my first time exploring the rainforest with a UV lamp, also known as a blacklight, in search of any sign of flora, fauna, or microbe that might be visible at night. The harsh conditions of the white-sand forests create selective pressures that lead to high levels of endemism—meaning the wildlife that lives here is found nowhere else on Earth. If there was ever a place to discover new luminescent life in the Amazon, it was here. At the time of this expedition, the only scientifically reported glowing organisms found in the Amazon Rainforest were bioluminescent click beetles that emitted a fascinating orange glow during their larval stage.

Luminescence, the ability to glow in the dark without heat, has fascinated humankind since the beginning of time. This curiosity was first recorded in prehistoric cave paintings over 30,000 years ago that depicted auroras as celestial manifestations that brighten up the sky. Over the centuries, notable figures like Aristotle, Columbus, and Darwin documented glowing wildlife and oceans in their writings, integrating the poorly understood mystical presence into cultural and scientific narratives.

Historically, luminescence, from the cosmic to the microscopic, has been considered a supernatural phenomenon, sparking awe and fear in folklore and myths. In Greek tradition, glowing organisms were considered divine symbols of light attributed to the powers of the god Poseidon or his nymphs, suggesting they possessed a mysterious ability that humans lack. Meanwhile, in Japanese folklore, glowing fireflies were considered the souls of fallen warriors. In maritime tales, these luminescent beings were thought to guide sailors to safety, with Roman historian Titus Livius describing "the shores [. . .] so luminous with frequent fires" that the "sea was aflame."[1] Far from mere myth, this ancient fascination with luminescence as a navigational aid proved lifesaving for Apollo 13 astronaut James Lovell. While flying a jet from an aircraft carrier off the coast of Japan, there was a malfunction and all the cockpit lights went out. Disoriented in pitch darkness, Lovell noted a faint green glow in the water below. The bioluminescent algae, stirred by the turbulence created by the aircraft carrier's movement through the ocean, provided a shimmering trail. Lovell followed this glowing path back to the safety of the ship, guided by the "flames" of the sea.

It wasn't until the 19th century that the chemical and biological processes behind luminescence began to unfold. The chemist Raphaël Dubois created a glowing paste using mollusks and wood decay fungi, proposing that "bioluminescence" was a chemical reaction, and coining the terms "luciferin" and "luciferase"—derived from the Latin *lucem ferre* (light-bearer)—for the substances responsible for producing light in organisms. As other scientists built on the work of Dubois, luciferin and luciferase became key in advancing medical and biotechnological research, from facilitating diagnoses and tracking disease, to enabling pollution detection in the environment, and microbial contamination identification in food and drinks.

Luminescence encompasses different mechanisms that enable wildlife to glow, including bioluminescence and biofluorescence. In bioluminescence, the glowing results from an internal chemical reaction that actively produces light, like fireflies that emit a captivating bright yellow glow or ocean plankton that decorate waves with the turquoise blue that was once referred to as "vivid flames" by Charles Darwin.[2] Biofluorescence, on the other hand, is the physical process through which an organism absorbs UV light (invisible to the naked eye) and re-emits it as visible light, thus appearing to glow. This process was first scientifically documented in the early 1960s. Japanese scientist Osamu Shimomura was studying a bioluminescent jellyfish when he discovered a specific protein, GFP, that would glow only when exposed to UV light. This revolutionary discovery, which won him the Nobel Prize in 2008, allowed experts to track gene activity and inspired the development of new medical treatments, among other biotechnological breakthroughs.

The ability for wildlife to luminesce has been referred to as a "foreign language" or "language of light" that we have yet to decode. These natural phenomena serve multiple roles in biology, including mating, communicating, facilitating camouflage, species recognition, and hunting, thus having deep ecological significance. While some instances of fluorescence might not serve a direct biological purpose, they can still be advantageous when applied in technological fields. For example, although the fluorescence of human teeth under UV light—caused by minerals and organic components—does not provide a survival advantage, it assists dentists in detecting bacterial infections and tooth decay, thus speeding up diagnosis and treatment.

Today, luminescence continues to create an aura of surprise, feeding human curiosity everywhere across the globe. From bucket-list travels to witness aurora phenomena to deep dives in the ocean to appreciate bioluminescence, we find glowing beauty all over our planet.

As we continued trekking, I noticed something moving under a pile of crunchy brown leaves. In a split second, a perfectly camouflaged frog bearing dark fungal spots jumped quick and high into the cleared ground, landing on thick white sand. A swift sweep of my UV torch revealed no biofluorescence from the frog before it croaked away into the darkness. No disappointment: we were only just starting. My years of training in science have taught me that sometimes your most intriguing discovery comes at the last moment, after hundreds if not thousands of attempts. During my Ph.D., I had to run over 1,000 biochemical reactions over the course of several months until, at last, I detected a positive result one Friday afternoon: a result that became the backbone of my 300-page dissertation. The trials were not failed attempts—they were stepping-stones to what ultimately allowed me to modify a microbe to generate antibiotics that were new to science.

As I adjusted my headlamp, which seemed to be dimming in power, I heard a loud gasp from some of our team members ahead. Aware of some of the perils that these lands harbor, including deadly venomous reptiles, I rushed up the hill, hearing endless chirping and squeaking all around. I could barely see anything in front of me until Luis, a field biologist, and Raul, a local student, pointed their headlamps at the thin tree trunk right in front of us. We stood in awe, witnessing one of the most dangerous creatures of the rainforest: a bright green and yellow round caterpillar.

Although not luminescent, the caterpillar bore the most vivid colors, adorned with brown lines that ran along its length, and dozens of orange spines. Each spine was further decorated with tens of needle-like turquoise tips radiating in all directions, creating a fascinating contrast with its green body and the pitch-black vastness of the rainforest.

– Doctora, *the spikes of this caterpillar cause immediate internal hemorrhage* – Luis said, his voice echoing in the rainforest. *Last year, a new species was discovered in a very traumatic way. An international traveler rested her arm on a tree, landing right on top of the caterpillar.*

A collective shudder rippled through the group.

– Wrong place at the wrong time – he continued, shaking his head. *She bled internally for over a week, but luckily, she recovered well, or so I heard. They usually live in a group, but this one is alone for some reason... In fact, this might be a new species...*

This was the first time I had come face to face with the infamous nocturnal "killer caterpillar."

Its caricature-like appearance belied the reality of its power as one of the world's deadliest creatures. The caterpillar stores its venom in its multiple colorful spikes. If stung, one will suffer from internal hemorrhage within hours that may last for weeks. Previous cases have reported that the victim can bleed from the eyes, eventually leading to compression of the brain and potential death. In fact, a specific species of this genus, *Lonomia obliqua*, was deemed the most venomous caterpillar worldwide by Guinness World Records. An antiserum was recently developed in Brazil to reverse the coagulation disorders caused by the venom, allowing the patient to recover quickly. The killer caterpillar's venom has been the subject of many medical studies due to its potent anti-clotting agents. Despite their dramatic nickname, these caterpillars don't seek to cause harm, but if you cause them stress and are close enough to touch, you may be a victim of their venomous spikes.

Even though the caterpillar wasn't biofluorescent, this find was still valuable, leading us closer to what we were looking for and indicating that we were approaching areas rich in biodiversity. We pressed forward with the sounds of the rainforest as alive as ever around us. Carefully, we crossed a narrow stream, balancing on a slippery tree trunk with our equipment lifted high. Although the water levels were low, we were wary of hidden dangers below.

For an hour, we trekked on, swinging the UV torch to detect any glowing species, but without success. Eventually, we reached a cliff edge. Walking in single file, we navigated the quartz-rich white sands, which sank beneath our boots, a contrast to the usual brown forest floor. To our right, a sheer drop loomed, with the dense rainforest canopy far below stretching as far as the moonlight shone.

After descending a hill, the cliff ended and we entered a denser path. Tangled roots, lianas, and fallen tree trunks were on the ground everywhere we looked, turning our journey into an advanced-level game of *Mario Bros.*, where we hopped, dodged, and weaved through the obstacle course of the rainforest.

Atop each liana and trunk, hundreds of unique fungal and microbial lives flourished: vibrant orange-yellow fungi in large clusters, lush green patches of lichen shaped in round medallions, and delicate white mushrooms that resembled flower petals, each dotted with dozens of black speckles. And as I continued swaying the blacklight torch, the first hint of Amazonian luminescent life emerged.

Everyone fell silent before breaking into an exciting clamor.

We all rushed to turn off our lights and welcome the infinite darkness surrounding us. My stomach muscles squeezed in excitement.

I switched the blacklight on again.

The powerful UV torch illuminated a section of the forest ahead of us in a soothing purple. Although pure UV light is invisible to the naked eye, our torch also emitted some visible violet light, creating the purple hue. But amid the rainforest darkness and the faint purple tone, a vibrant turquoise glow shone from a single tree trunk.

Upon closer examination, we learned that this turquoise light emanated not from the tree itself, but from lichen—a symbiotic relationship between fungi and algae—coating the trunk. This lichen, glowing in a mesmerizing turquoise, reached from the forest floor to over a meter high. Its luminescence casted an enchanting aura, accentuating the hidden textures of the lichen so that it resembled crumpled Origami paper meticulously glued to the surface of the tree.

It is still not fully understood why one of the most resilient life forms on our planet—lichen—fluoresces under UV light. A widely accepted theory is that its ability to glow acts like a sunscreen, helping the lichen to deflect harmful UV rays and protecting its internal structures from UV-induced damage. In some instances, the glowing components may simply be waste products of the lichen's daily life, which neither interfere with its processes nor add a beneficial advantage. Due to their ability to adapt to extreme conditions with limited resources, lichens are great indicators of ecosystem health. Thus, studying their glowing abilities may help us to develop strategies for environmental monitoring of pollution and climate change.

Biofluorescence, the process through which this lichen was glowing, was first documented in the 19th century. Physicist Sir David Brewster observed the mineral "fluorite" emitting a blue glow when exposed to UV light, and coined the term "fluorescence." Later, polymath Sir John Herschel discovered that quinine sulfate—derived from the bark of the Amazonian cinchona tree (see page 206) and a key ingredient in tonic water—emitted a blue fluorescence when exposed to sunlight. In fact, this phenomenon is the basis for a common party trick where a glass of gin and tonic emits a striking blue glow under UV light. But other than offering a fun visual, the question remains: why do some animals glow?

Over the years, biofluorescence has been observed in a variety of organisms in a spectrum of colors, including blues, greens, and reds.

Initially, attention was largely centered on marine animals, as the lack of light in the deep ocean was recognized as an evolutionary pressure for these creatures to develop light-emitting mechanisms for survival. This led to the discovery of biofluorescence in corals, turtles, and over 200 species of fish and sharks. However, it has only been within the last two decades that scientists have intentionally expanded this quest to land animals, discovering biofluorescence in flying squirrels that emit a stark pink glow, and platypuses that shine with a purple-green hue.

Biofluorescence is an area of ongoing discovery, with a wide variety of creatures found to glow under UV light, including fungi, owls, birds, frogs, spiders, chameleons, snakes, and even the fur of the thylacine (also known as the Tasmanian tiger), the world's largest extinct marsupial. This range of natural luminescent life, from aquatic and terrestrial to aerial, fills our world with both explicable and inexplicable glows that color our view of the planet. It inspires new avenues through which to appreciate and protect wildlife wherever we find it. Although much is still left to discover and understand, biofluorescence undoubtedly inspires a sense of wonder at the otherworldliness of nature in all of us.

Wondering what secrets this Amazonian lichen may hold, we paused to appreciate and document our finding, reinvigorated by our discovery and eager to continue exploring the nocturnal wildlife of the rainforest. Under white light, the lichen appeared olive green, identical to the lichens on at least four nearby tree trunks, with no detectable differences to the naked eye. This observation further intrigued me: why did only one collection of lichen on a specific tree glow, while others faded into the darkness?

This led me to consider how the environment contributes to the ability of organisms to luminesce. It is plausible that a unique set of conditions must be met in order for wildlife to glow. For example, bioluminescent organisms, like fireflies, depend critically on the availability of oxygen or cofactors. Additionally, variables like temperature, pH, and salinity may also influence the likelihood of light production. In the case of biofluorescent life forms (like the lichen emitting a turquoise glow under UV light), specific wavelengths of light are essential, with variations in time of day playing a crucial role. Several other factors impact the ability of wildlife to glow, whether through bioluminescence or biofluorescence, including genetic variation, developmental stages, nutrient availability, stress and pollution, and interactions with other organisms. These elements highlight the adaptive significance of luminescence in the natural world. Perhaps the glowing lichen before us harbored genetic variations that enabled it to emit light, while its counterparts remained hidden in the pitch black.

We advanced through the dark, with only one headlamp and our UV torches. With minimal visible light, the sounds of the jungle seemed to intensify: low- and high-pitched tunes, flutters, and calls that resonated in our ear drums. The aromas also grew stronger, including the faint but distinctive peppery fragrance of matico plants, a sacred ingredient in my grandmother's traditional recipes for wound-healing and asthma. The sounds and scents accompanied us as we pushed forward, expanding the boundaries of our own curiosity and drawing us closer to new scientific discoveries.

As we neared another stream, we activated all our headlamps to provide a better view and prevent anyone from slipping in the darkness. The scents of crisp water and moist soil filled the air and our lungs, while the moonlight filtered through the canopy, reflecting the silhouettes of the surrounding flora.

– *A glass frog!* exclaimed Luis, as he examined long, thick leaves nearby.

We approached as Luis pointed to the heart of the plant, where broad and elongated leaves radiated out from the central veins. Nestled right in the center was a tiny frog, protected by the microhabitat and canopy provided by the leaves. Its mostly translucent body, subtly tinged green with small darker green spots, mirrored the leaves, allowing it to seamlessly blend into its surroundings. Its large, black expressive eyes were delineated by a golden rim, adding to its ethereal appearance.

In a reflex motion, I turned on my UV light. That's when we encountered the unexpected: a fluorescent pattern that none of us could have predicted.

The frog's eyes—and only the eyes—were emitting a bold greenish-turquoise glow.

Thinking my brain was playing tricks on me, I turned on the headlamp to see the animal in white light again. I squinted, trying to detect any subtle physiological detail on the frog's eyes (lines, shapes, colors) or signs of microbial infection (redness, swelling, discharge) that might hint at an explanation. However, the eyes revealed nothing but deep, impenetrable darkness like a starless night sky. The frog remained immobile, staring back.

The world's first biofluorescent frog was found in South America in 2017. At the Natural Sciences Museum in Buenos Aires, Argentina, scientists conducting a study on the translucent polka-dot tree frog (*Hypsiboas punctatus*) shone UV light on the specimen to analyze tissue samples. To their surprise, they discovered that the entire body of the frog, except for the eyes, emitted a vivid green glow. Although the specimen in the study was collected from Argentina, this species is widely distributed across South America, including in parts of the Amazon Basin. Subsequent research revealed that dozens of other amphibians, including other types of frog and salamanders, could also glow under UV light. However, the study did not specify the precise location from which the animals were collected, nor did it detail which specific body parts glowed, beyond noting that the ventral and dorsal surfaces of frogs were the most commonly glowing areas. The findings suggested that the fluorescence enhances the visibility of these frogs under the low-light conditions of their environment, suggesting that their luminescence may be essential in communication, finding mates, or avoiding predators.

In pitch darkness, I turned on the UV lamp again, to see once more that vibrant and undeniably striking greenish-turquoise glow. The frog croaked softly, its tiny body expanding subtly as it respired.

In awe and lost for words, we were observing an unexpected layer of beauty in the rainforest that was unfamiliar to us all. It felt as if a small UV torch had opened a door to our own living Narnia, allowing us to directly appreciate the palpable, breathing jungle telling a story in colors and lights right before us. Like a show of fireworks, with sparks and bursts flying back and forth, these displays transported secret signals and messages that remain inadvertently invisible to the human eye, allowing wildlife to survive throughout time. These are signals and messages I hope we can someday understand. Tapping into the "language of light" might allow us to rediscover the wonders that live on planet Earth—wonders that are worth fighting for.

This sense of wonder brought back memories of my first encounter with fireflies in North America.

On a humid summer night in Tennessee when I was 18, having just moved from Perú to start my science degrees, I watched dozens of fireflies pulsing with light, creating a magical ambiance in the air, before they disappeared into the dark. I wondered if this was how poets from ancient China had felt when they described fireflies in their writings as natural marvels—beacons of light piercing the loneliness of the night, each a fleeting moment embodying longing. In the words of Tu Fu, a great poet and official of the Tang Dynasty: "Flares of color against the dark... I try to read their code... will I be here next year to watch them?"

At the time of our observation, there were no published scientific articles indicating that frogs collected from the Amazon Rainforest exhibited fluorescence, particularly not solely from the eyes. Typically, biofluorescent animals on our planet emit light from their entire bodies, enhancing visibility for camouflage or communication; the glow usually forms part of an extensive pattern rather than being limited to the eyes. A rare exception, like the tiny frog before us, are flashlight fish (*Anomalopidae*). These marine creatures live up to 400 meters (1,300 feet) below the surface of our oceans, where light is minimal or absent. They possess organs just below their eyes that house bioluminescent bacteria, allowing the fish to "blink" their turquoise lights on and off—much like a light switch at home. In the profound darkness of the deep ocean, they resemble floating, glowing turquoise orbs, looking like the inhabitants of an alien world. This bioluminescence allows them to forage in the dark waters, attracting plankton and other small organisms that are drawn to the light in a very dark environment. Their ability to toggle this light also helps disorient predators. This is a remarkable example of wildlife adaptation to extreme environments, where, much like the densest corners of the Amazon, shadowed by lush canopies, light is scarce.

In July 2023, a research article was published that surveyed the Amazon Rainforest to identify frogs and toads capable of biofluorescence at multiple wavelengths, mimicking the natural light found in their environment. Of the 151 species studied, three were found to exhibit biofluorescence from their eyes (none of which were the species we saw). Remarkably, only one species, *Boana calcarata*, emitted a bright green glow exclusively from the eyes. Therefore, in the middle of the white-sand forests of the Peruvian Amazon, we had identified one of only two currently known species of Amazonian frogs that glow solely from the eyes.

As we continued to explore our surroundings through this new lens, we discovered more sparkles of color radiating from the rainforest. Glowing microbes adorned leaves in bright blues and greens, their colors scattered throughout the forest without any obvious connection to the flora hosting these luminescent creatures, adding a layer of mysticism that invited us to explore deeper. Walking amid the glowing Amazon, I was reminded of German naturalist Georg Eberhard Rumphius, who in the late 17th century documented how indigenous people in Indonesia used bioluminescent fungi to illuminate their paths through dense, dark forests. Here, instead, our UV torches revealed specks of light in the wild that guided our footsteps.

As we continued surveying our surroundings, my husband alerted us to a glowing scorpion that had emerged from the shadows, previously hidden among the foliage. It gave off an ethereal glow under the blacklight. Every segment of the scorpion's body, from its pincers to its tail, was turned into a living jewel as it shone with a vibrant blue-green biofluorescence that contrasted sharply with the white sands and dark plant matter. Although I was aware that all scorpions glow under UV light, this was my first time witnessing it. As we watched, mesmerized, the scorpion quickly hid itself beneath a thick mass of dead leaves.

All scorpions glow in the dark—whether you are in the Amazon, Europe, Asia, or anywhere else. A substance within the scorpion's exoskeleton called beta-carboline absorbs UV light and re-emits it as a visible turquoise light. Their fluorescence is so persistent that scorpions will continue to glow even when preserved in alcohol for scientific purposes, causing the alcohol itself to glow. While it is not fully understood why scorpions glow, the most widely accepted theory suggests that by absorbing UV light—particularly on moonlit nights—their bodies send signals to their brains. This helps them make decisions about whether to stay hidden or to move around, with a preference for scouting for food and shelter under the cover of darkness.

A few seconds after we saw the scorpion, a coral snake (*Elapidae*) emerged from below the leaves, hissing and sliding its way across our path. Striking patterns of bright red and luminous black bands encircled its body, the very definition of warning colors, as this snake has a highly venomous bite. Coral snakes are typically non-aggressive and tend to avoid human contact. I remained still, watching its magnificent scales reflect the lights from our headlamps until it disappeared out of sight. With laser-like focus, I scanned the forest floor for any sign of other snakes lurking nearby. Unlike the evasive coral snake, not all serpents in these lands shy away from human interaction.

This area is home to one of the most enigmatic species in the Amazon: the *shushupe* (*Lachesis muta*). According to local lore, this is the only snake known to actively chase after humans in the jungle. Recognized as the longest viper in the world, reaching over 3.5 meters (11½ feet) in length, the *shushupe* is one of the most feared and respected reptiles in the rainforest. Its genus name, *Lachesis*, derives from the Greek goddess Lachesis, who was said to be in charge of determining the length of human life—a reference to the potency of the snake's venom. Also known as the Amazon Bushmaster, its light brown or reddish-brown tones, marked with darker colorations, make it well-camouflaged in the leaf litter of the forest floor. Primarily nocturnal, the *shushupe* prefers humid and dense areas where it can hunt small mammals and birds, and it plays a key role in controlling rodent population. Habitat destruction and hunting are currently threatening the species, with conservation efforts becoming more crucial than ever.

A few days before I embarked on the expedition, I ran into Avi, the remarkable biologist and local guide with vast knowledge of the rainforest, especially the wildlife that wakes up at night. She shared with me one of the most harrowing encounters I have ever heard between a human and a *shushupe* snake.

– *I was trekking the rainforest past midnight, exactly where you are heading to,* doctora, *when I ran into the* shushupe. *I was taking an American lady on a tour to appreciate the nocturnal wildlife* – she began. *I saw that the* shushupe *was protecting her eggs. We had accidentally stepped into her nesting site, scaring her.*

Avi stood up from the couch to better describe what happened next.

– *She coiled her body around, her head standing up and tall. I immediately pushed the American lady behind me, right on time, as the* shushupe *jumped forward and bit my rubber boots.*

She portrayed the fast reaction of the snake by quickly extending her arm toward my face, making me jump in surprise.

– *What . . . what did you do?* I mumbled.

– *I stood still. I knew she was defying me. She was protecting her eggs.*

So I didn't move an inch, and stared right into her golden eyes. I wanted her to know that I wasn't there to attack her. She moved her large body, coiled again, and repeated to charge. And another time after that. She bit my rubber boots a total of three times. I knew my boots were thick and her bites wouldn't get to me, so I held my ground.

I watched Avi, mesmerized.

– Finally, she stared at me for a minute, flicking her tongue, before slowly retreating. Once she was a few meters away, I slowly walked backwards, still staring at her. I removed my sweater and threw it a few feet in front of me—then off we went.

Amazonian legend says that if you offer a piece of your clothing to the *shushupe*, the snake will take it as an offering and leave you alone. I think Avi may be one of the bravest women I have ever met.

Recognizing that we were in territory ideal for snakes, we proceeded with increased caution, carefully checking any pile of leaves and forest debris. The path became increasingly treacherous as mud slicked the ground, our leg muscles tensing to maintain balance and avoid stumbling into trees or leaf mounds that might harbor killer caterpillars or *shushupe* snakes.

We kept trekking for another couple of miles, without any major observations or discoveries. Feeling already fortunate to have encountered such a variety of nocturnal wildlife in these lands, including observing the killer caterpillar and discovering a frog that glowed exclusively from its eyes, we decided to call it a night. We veered right and followed a smaller trail that would take us out of the deep dark forest and into the cleared grounds that preceded camp.

But the Amazon had other plans.

While we chatted energetically, just joking around, we continued to sweep our UV torches, purely due to the diligence of scientific duty. And there was one final life form that invited us into its ancient secret language of light.

To my left, among leaves twice the size of my hands, stood a giant grasshopper that gave off a lime-green glow from its head and legs, while the rest of its leaf-like body remained in the dark, including its eyes.

Testing the blacklight a few times confirmed that the fluorescence came solely from the head and its six legs. The glow might play a role in mating behavior and territorial displays, or just offer camouflage by making it harder for predators to recognize their shape in the low-light conditions of heavy Amazonian canopy. It is also possible that the concentration of biofluorescence on its head and legs results from structural processes during the grasshopper's growth, as has been suggested in other animals such as birds, where the fluorescence accumulates on the feathers.

We all looked at each other, amazed. Our observations expanded what is known about nocturnal animals in the Peruvian Amazon and their capacity for biofluorescence, suggesting a complex mix of behavioral interactions that have long remained a mystery. The possibilities of discovery through luminescence felt endless. Despite being a group of experienced explorers, looking at wildlife through a new lens had taken us to the verge of what felt like a new era of exploration.

During our night expedition, no samples were collected. Instead, wildlife was observed in its natural habitat, igniting our desire to return and continue learning from the Amazon at night. We didn't find trees that glow like bright blue bulbs, but we had certainly stepped into the extraordinary—an Amazonia that I had never witnessed before, even though I've been traveling to the rainforest since I was one. It was a utopia of color, a glow-in-the-dark party of life, where every flicker of light evoked a deep and unshakable sense of curiosity, beauty, and hope.

Walking through the luminescent rainforest, we felt like astronauts discovering new sources of light in space, except our feet were firmly grounded on Earth. Each glowing life form we encountered in the Amazon that night was enveloped in a spellbinding aura of mysticism, deepening our understanding of the spiritualism associated with luminescence throughout history.

Nature has perfected ways of communication since the dawn of our planet, long before human civilization began. From the depths of the ocean to the dark corners of the rainforest, minimal light has fostered unique environments that have given rise to languages among flora, fauna, and microbes that remain largely undeciphered. Nature is communicating through light, speaking in a luminescent dialect we are only just beginning to understand. Tapping in to this ancient dialogue could inspire our society to forge a deeper connection with our planet, our Pachamama.

We do not need to travel to the edges of the Earth to appreciate the mysticism of our wildlife. Sometimes, all it takes is a new method—like that offered by luminescence—to unlock a fascinating way of understanding the natural world and the pressures it faces. A new age of adventure and exploration awaits, within the grasp of anyone eager to deepen their appreciation for species and ecosystems that may have been previously overlooked.

By studying creatures that glow in the dark, we are delving into the essence of human curiosity, reconnecting with the divine nature our ancestors revered, and charting new paths for appreciating our natural world. Just as Aristotle marveled at the "fire in the sea,"[3] we stand awestruck by the fireworks of light in the rainforest, a language that surpasses our wildest imagination, reminding us that everything in nature is profoundly beautiful and worthy of admiration.

EPILOGUE

I sat in discomfort. Mosquitoes were feasting on my neck. The microphone cord that had been carefully placed under my shirt was itching. The wind ruffled my long black hair, and I was resisting the impulse to sneeze. A drop of sweat mixed with humidity was stinging my left eye, but I didn't dare wipe it off in case I smudged my mascara. After all, this was my first time on daytime TV, streaming live from the jungle.

CBS had flown in two of their top reporters and camera guys to join us in the heart of the rainforest for an interview about our work on the ground. It had all happened very quickly; a cover article in the *New York Times* about our conservation efforts had led to interviews, videos, and a series of social media posts. Through a single tweet, I connected with a Latin American editor, and in less than three weeks we were traversing the heart of the Peruvian Amazon. We trekked deep into the forest, snaking through hills, bushes, and streams in search of the stingless beehives that we aimed to preserve through our conservation efforts.

– *Why do you do it?* asked the reporter, bringing me back to the present. His cheeks were bright red, his face dripping with sweat, eyes oozing curiosity. Deepening creases around his eyebrows and the slight right tilt of his neck revealed his genuine bewilderment at seeing the intense work that took place behind the scenes. In his mind, I could be anywhere else.

I stared back. I could feel the group of people around us listening in. Children were giggling at the camera studio we had improvised in the Amazon, their backyard. Our indigenous partners and local scientists stood proudly having just finished their own interviews. I was the last one to speak.

Why do I do it?

Our rainforest was in flames. Smoke emanated from tree canopies into the sky because of devastating destruction. Acres of land and water were being stripped of their beauty. Obliterating floods and calamitous droughts were recorded across the forest. Animals squeaked in horror. People were bathing in rivers contaminated by oil spills, mercury, and plastic. It was clear that the relationship between humanity and nature was broken.

Yet with my experience of the rainforest, I witnessed a different story.

Just days earlier, we had been gathered around a traditional roaring fire in the heart of the Ashaninka community, where we joined a ritual of dance and song as the twilight gave way to night, the flames and sparks illuminating the dark jungle. The scent of burning wood intertwined with the forest's earthy aromas, awakening a primal sensation of safety and calmness. Behind us, the *Avireri* mountains stood majestic and protective, their breathtaking silhouette cast by moonlight that filtered through the dense canopy, evoking a sense of magic.

The *Apus* began singing and playing flutes and drums, their music echoing into the forest and blending with the symphony of sounds. In their traditional language, they shared stories of origin—how nature and people came to be—followed by songs of gratitude for every spirit and life form that guided us: human, plant, and animal. The women began singing high-pitched notes, their dresses adorned with sewn-in seeds that rustled like rain. Their words expressed longing, asking the rainforest to breathe, to fight, to stay alive. They danced fluidly in hypnotizing waves that reminded me of trees swaying in the wind. The flames drew shadows that danced alongside them like the spirits of their ancestors.

Suddenly, everyone burst into laughter—children included—clapping, hugging, and rejoicing in the shared happiness of a ritual where every voice was heard. *Living beautifully.* My eyes glistened brightly as joy and electricity ran through my spine and radiated through every corner of my body. These were the ancient ways in which Amazonians recalibrated their minds. Through dancing and singing, heart and spirit, they prepared to stand up and try again, inspiring me to fight ever harder.

I looked at the reporter and answered with a smile.

– It's our home.

The Amazon brings us back to our roots: to the simple yet profound truth that we are an extension of nature and beauty. It teaches us strength and resilience in the face of change. It shows us that generosity overpowers unfairness. The rainforest fills us with the courage to tap in to our natural instincts of curiosity and awareness, allowing us to make discoveries and illuminate answers that have sat silently, waiting to be found. It motivates us to push the boundaries of what is known and what is comfortable, and to embrace growth with laughter and joy.

I don't know where we will be in the future, but my journeys through the Amazon have taught me to embrace uncertainty with calm, control, and order, just as Sergio once told me; to surrender the ego, take courage, and accept that not all is doomed. The journey hasn't been easy, and there is still so much left to do as my quest to elevate indigenous knowledge within modern science goes on. These pages capture many stories and adventures, but countless more untold narratives still await.

I dream that these pages invite us to explore a different mindset, an indigenous cosmovision that will inspire us to forge a future that we can be proud of—a reality I have already found in the deepest corners of the jungle. The ancient wisdom of our ancestors teaches us that nature and people are a singular, intertwined entity, and through reconnection, there is a restored joy and balance.

Just like the endless colors, textures, and scents we find in the wild, there are infinite ways to live beautifully. To live beautifully is to reconnect. To reconnect is to love. And to love is to protect our Pachamama.

ACKNOWLEDGMENTS

When writing, I feel free. Before dance, astronomy, and science . . . writing was my first love. Since I was five, I've written everywhere I could physically reach—journals, computers, floors, and walls (sorry, Mom). I figured I loved literature so much that I would eventually come back to it, and so I chose to pursue science first. Truth be told, the idea of someone reading something I wrote terrified me since an English professor told me my writing was no good. It feels raw and vulnerable, even more so than trekking through the rainforest in pitch darkness.

Thank you to my agents, Adrian and Alice, for taking that first call, for believing in my work and character, and for seamlessly communicating my vision. To my editor, Jessica—thank you for giving me my first chance to become an author, and for your impeccable attention to detail, patience, and creativity in bringing this book to life. Thank you, Rimsha, for fine-tuning the wording into the creation it became. And thank you to the Octopus Books family for making a childhood dream come true.

To my husband, Chris, who first encouraged me to submit my book ideas, thank you for developing the backbone of this book with me,

ACKNOWLEDGMENTS

for reading draft after draft, and for cooking dinner every day while I fired away at my writing day and night. Thank you for uplifting me with laughter and joy, along with our tiny puppy Xavi, whenever I was filled with doubts throughout the process.

Thank you to my entire family—cousins, uncles, and aunts—for helping me remember the past, for embracing my wild adventures, and for making me feel like time hasn't passed whenever I return home. And to our Leona, who watches us from the sky, thank you for dancing with us to every line.

Thank you to my father, who instilled in me a passion for science, making me wonder about genetics at the age of six. Sitting down with you, drinking a lucuma smoothie, and discussing the pharmacopeia sections of this book is one of my favorite memories.

Thank you to my mother, who taught me how to sing to flowers and encouraged me to fly as high as eagles. You showed me how to embrace mistakes and navigate the unknown with kindness and a smile. Traveling across the Amazon with you by my side while completing this book was a poetic experience.

To my grandmother, the ultimate doctor, my guiding star in the dark—thank you. Words are insufficient to express my gratitude for you and your existence. I'm in awe of your wisdom and strength. From a remote corner of Perú to libraries worldwide, this win is yours as much as it is mine.

And although I mentioned it at the start, I must say thank you again to all the *Apus* and mentors who have enriched my journey and spirit along the way. Without your generosity, spirituality, knowledge, and hard work there would be no story to tell. Thank you for offering a kind smile and a helping hand, and may our work resonate for generations to come.

ACKNOWLEDGMENTS

And last but certainly not least, thank you, Pachamama. Thank you to the mountains for keeping me safe and thank you to the animals and plants for guiding me in times of distress. It is with the utmost respect that I enter your territory and share your tales and wisdom. For my future children and the children of their children, I hope to make you proud.

PICTURE CREDITS

Page 1 (above): Luis Reyes
Page 1 (below): Stephanie King
Page 2 (above): Lucas Fiat
Page 2 (below): Carlos Espinoza
Page 3 (above): Rosa V. Espinoza
Page 3 (below): Rosa V. Espinoza
Page 4 (above): Brenda Rivas
Page 4 (below): Freya Park
Page 5 (above left): Yuri Hooker
Page 5 (above right): Chris Perry
Page 5 (below): Ana Elisa Sotelo for *National Geographic*
Page 6 (above left): Luis Garcia Solsol
Page 6 (above right): Rosa V. Espinoza
Page 6 (below): Maria Espinoza
Page 7 (above): Luis Garcia Solsol
Page 7 (below): Will Stoll
Page 8 (above): Luis Garcia Solsol
Page 8 (below): Gustavo Carrasco

ENDNOTES

Introduction

1. The Pachamama is an Andean goddess—"Pacha" meaning Earth (or cosmos) and "Mama" meaning Mother. She reflects an indigenous concept of time and space, aligned with nature's rhythms. She is also known as the goddess of reciprocity and duality. In many Amazonian-Andean cultures, the Pachamama is thought of as a maternal feminine figure, and perceived as having nurturing qualities and being a provider and a protector.
2. Pisco, the national spirit drink of Perú, is a colorless grape brandy prepared with traditional distillation methods. It typically has an alcohol content of 40–50 per cent.

Chapter 1: Pachamama

1. Endemic to Perú, the rare yellow-tailed woolly monkey (*Oreonax flavicauda*) has been observed engaging in opportunistic foraging, sometimes taking food from neighbors or people. This species primarily feeds on fruits, leaves, and flowers, and is considered critically endangered by the International Union for Conservation of Nature (IUCN). Their main threats include deforestation and illegal animal trade.
2. When preparing the Amazonian drink *masato*, local women chew the yucca to initiate the process of fermentation. The naturally occurring bacteria in their saliva convert the starches in the yucca into carbon dioxide and alcohol.

ENDNOTES

Chapter 2: A River That Boils

1. Fire ants are known for delivering a burning sting when defending themselves. If stung, you first feel a sharp, acute pain as they inject the venom into the skin. Then, as the alkaloid poison disperses, you feel a burning sensation that extends far from the sting site, leaving a prickling red bump. Although the burning and intense pain only last a few minutes, some fire ants have the ability to sting you repeatedly.
2. When thermophiles thrive in temperatures above 80°C (176°F), they are known as "hyperthermophiles."

Chapter 3: Life in the Water

1. The phrase "as a subject of rights with intrinsic value" reflects the core principle of the Rights of Nature movement, which recognizes that nature—ecosystems, species, and natural entities—possesses inherent value independent of its utility to humans. This legal philosophy asserts that nature has the right to exist, thrive, and regenerate, and that these rights should be legally recognized and protected. By granting nature legal personhood, it shifts from being viewed merely as property to being a rights-bearing entity that can be defended in court.

Chapter 4: Evil Spirits

1. Also known as the Peruvian Inca Orchid, this ancient breed of dog is characterized by its lack of hair and dark, smooth skin. They are known for their loyalty and intelligence, making them great guard dogs. Many artifacts depicting this dog have been found in ancient tombs. The breed is nowadays recognized as a national treasure in Perú.
2. Bullet ants (*Paraponera clavata*, traditionally known as *isulas*) are some of the largest ants in the world, measuring up to 2.5 centimeters (1 inch) in length. They have black, shiny bodies and, although primarily found in the ground, they are also known for climbing trees. They are found across Central and South America, and they are renowned for their excruciating stings, which are often compared to the sensation of being shot—hence the name. The agonizing pain they cause is primarily due to a neurotoxin they produce called poneratoxin. This chemical affects voltage-dependent sodium-ion channels in the nervous system, leading to pain, shaking, sweating, and even heart palpitations. The pain can last for several hours.

3. *Huayruro* (*Ormosia* sp.) is a seed derived from an evergreen tree abundant in various regions of the Amazon. Recognized for its distinctive colors and attributes, it has been widely utilized in traditional jewelry, fashioned into bracelets, necklaces, and other ornaments. Occasionally, these seeds have served as natural insecticides for crops and have been cautiously employed in treating bacterial or fungal skin infections.
4. Although they are completely safe to handle, these seeds contain toxic alkaloids that are poisonous if ingested.
5. Biocultural heritage is a concept recognized by the United Nations that emerged within the fields of conservation, ecology, and anthropology. It refers to knowledge that brings together people, language, and culture, and their relationship to nature and the environment.
6. A snake's forked tongue is a remarkable proof of nature's adaptation methods. The two prongs of the forked tongue allow the snake to collect scents from two different locations at the same time, creating a precise map of the scent's source. This ability is key in helping snakes to follow scent trails, detect predators, and locate potential mates.
7. Buffer zones in the Amazon Rainforest serve as areas surrounding protected zones, acting as transition spaces between human activities and the core conservation areas. While they are considered essential for effective conservation, buffer zones often face exclusion from funding and conservation initiatives. This oversight can lead to increased destructive practices in the jungle.
8. A significant portion of Peruvians, including my mother, are Catholic. My father is a non-denominational Christian, my grandmother is a Jehovah's Witness, and some in my family are spiritualists. Our Amazonian ancestors were animists, believing that everything in nature possesses a spirit or soul. These beliefs are not tied to a specific religion, but are rather rooted in a deep connection with our natural world. As an adult, I appreciate the beauty of having grown up with multiple belief systems and worldviews, and deeply respect each one of them.

Chapter 5: Stingless Bees

1. Herodotus, *Histories*: *Book IV*.
2. Stinging bees, including honeybees and bumblebees, are estimated to have evolved around 30–40 million years ago. They belong to the family *Apidae*.

3. An entomologist from the University of California, Davis, developed a sex pheromone known as Queen Mandibular Pheromone (QMP). This particular complex honeybee pheromone is secreted by the queen herself and it plays a crucial role in fostering a sense of belonging to the queen among hive members. When shared with the hive, it serves to attract males to the unmated queen.
4. The Amazon's humid environment leads to higher water content in Amazonian plants. This contributes to raised moisture levels in the nectar. Consequently, this results in higher liquid levels in the honey produced by Amazonian stingless bees. The elevated moisture makes this honey less viscous compared to the honeybee honey that we purchase in a supermarket.

Chapter 6: Ayahuasca

1. *Muña* (*Minthostachys mollis*) is a woody medicinal shrub native to the Andes in Perú. This highly aromatic plant, with a mint-like flavor, is the smell of my childhood and makes for a delicious tea my grandmother will happily prepare fresh for any visitor stopping by.
2. Also known as *Piper aduncum, matico* is a flowering plant with a peppery aroma and powerful antiseptic properties, used in Amazonian traditional medicine.

Chapter 7: What Lurks Beneath the Murky Waters

1. Green anacondas, including those with blackish colorations, are regarded as the heaviest and among the longest snakes in the world. They can exceed 8 meters (over 26 feet) in length, and weigh more than 227 kilograms (500 pounds). Capable of preying on larger mammals, such as capybaras, caimans, and even jaguars, these reptiles are semi-aquatic and spend much of their lives in rivers and swamps. In fact, they ambush their prey from beneath the water's surface.
2. The Amazon is home to the world's most powerful electric eel, which can deliver a shock of 860 volts of electricity—the highest voltage discharge measured for any living animal. Many other eel species are found in the rainforest, each adapted to the complex characteristics of this basin.

Chapter 8: Living Fossils

1. The story of Pirarucu is well known among Amazonian communities and is shared among kids to serve as a lesson in humility and respect, symbolizing the gods' power and reminding us to uphold morality within the jungle.

2. Although not living fossils, parrots have a rich evolutionary history and ancient lineage, having undergone various specialized adaptations in order to thrive in the tropics. One of these adaptations is their vocal mimicry ability, which allows them to imitate sounds from their surroundings, including human voices. This skill helps parrots develop social bonds, deceive predators, and even blend into a landscape to avoid detection.

Chapter 9: Living in the Clouds

1. *Chimpa* derives from the Quechua language and refers to the act of "crossing;" these are platform infrastructures that unite towns and communities.

Chapter 10: Ancestral Civilizations

1. The *huito* tree is native to the Amazon Basin, with a straight trunk that often reaches up to 15 meters (50 feet) in height. The tree produces small, white fragrant flowers, and its fruits are similar in size and shape to small avocados. In addition to the tree's cultural and artistic significance, the fruit, leaves, and bark are traditionally used for their medicinal properties and for making natural dyes.
2. Morpho butterflies (*Morpho* sp.) are some of the most stunning and beautiful butterflies in the Amazon—at least, in my opinion—and are also among the largest butterflies in the world. With iridescent blue wings and a remarkable wingspan of up to 15 centimeters (6 inches), their tiny scales reflect sunlight, acting as a protective shield against predators—like flashing a light three times into the dark. They are important pollinators of native flora, and have contributed greatly to the scientific understanding of structural coloration in nature.
3. Today, we have approximately 7,000 living languages globally, of which 4,000 are indigenous. Unfortunately, over 40 per cent of languages are currently endangered and at risk of disappearing, including the Bora language.
4. Most ancient cultures lacked a formal concept of zero or used it as a placeholder, but the Maya recognized it as a numerical value as early as the 4th century CE. This marks one example of the early development of "zero," which played a critical role in mathematics and astronomy.
5. One of the earliest applications of LiDAR technology, first developed in the early 1960s, was in the Apollo 15 mission in 1971, when it was used to map the surface of the moon.
6. Unfortunately, nowadays, it is very difficult to find sections of rivers or springs that are free of contamination, putting the health of the consumer at great risk.

ENDNOTES

Chapter 11: In the Eyes of the Jaguar

1. Jaguars' jaws can exert a force of 1,500 to 2,000 psi (pounds per square inch). For comparison, domesticated house cats have a bite force of only about 70 psi.
2. The "world of the living" is also known as *Kay Pacha* in Quechua, and refers to the perceptible world that people, animals, and plants inhabit.
3. This "bee glue" is so sticky that Amazonians once repurposed it to construct their hunting arrows, musical instruments, handicrafts, and fishing tools. Amazonian communities knew that this resin could also be used to make candles to provide clear and long-lived light beyond the hours of the sun.
4. Officially, jaguars are classified as near-threatened by the International Union for the Conservation of Nature (IUCN). The species is protected nationally in almost every country it inhabits, and international trade in jaguars or their parts is banned globally under the Convention on International Trade in Endangered Species (CITES).
5. *Cuniculus* sp., also known as *paca* or *majás*, is a ground-dwelling rodent with dots and stripes on its sides, and a tiny tail that is barely visible. It is the sixth-largest rodent in the world, and feeds on flowers, insects, leaves, and fungi. It represents an important food supply for Amazonian people.

Chapter 12: Life that Glows in the Dark

1. Quoted in Latz, M. and Rohr, J., 2005, "Glowing with the flow." *Optics and Photonics News*, 16(10), pp. 40–45.
2. Darwin, C. (1845) *Journal of Researches into the Natural History and Geology of the Countries Visited during the Voyage of H.M.S. Beagle Round the World, under the Command of Capt. Fitz Roy, R.N.* John Murray, London.
3. Passage in "De Mundo" (Letter to Alexander the Great) by Aristotle. Quoted in Harvey, E.N., 1957, "A history of luminescence from the earliest times until 1900." *Memoirs of the American Philosophical Society*, 44.